SpringerBriefs in Applied Sciences and Technology

PoliMI SpringerBriefs

Series Editors

Barbara Pernici, DEIB, Politecnico di Milano, Milano, Italy
Stefano Della Torre, DABC, Politecnico di Milano, Milano, Italy
Bianca M. Colosimo, DMEC, Politecnico di Milano, Milano, Italy
Tiziano Faravelli, DCHEM, Politecnico di Milano, Milano, Italy
Roberto Paolucci, DICA, Politecnico di Milano, Milano, Italy
Silvia Piardi, Design, Politecnico di Milano, Milano, Italy
Gabriele Pasqui, DASTU, Politecnico di Milano, Milano, Italy

Springer, in cooperation with Politecnico di Milano, publishes the PoliMI Springer-Briefs, concise summaries of cutting-edge research and practical applications across a wide spectrum of fields. Featuring compact volumes of 50 to 125 (150 as a maximum) pages, the series covers a range of contents from professional to academic in the following research areas carried out at Politecnico:

- Aerospace Engineering
- Bioengineering
- Electrical Engineering
- Energy and Nuclear Science and Technology
- Environmental and Infrastructure Engineering
- Industrial Chemistry and Chemical Engineering
- Information Technology
- Management, Economics and Industrial Engineering
- Materials Engineering
- Mathematical Models and Methods in Engineering
- Mechanical Engineering
- Structural Seismic and Geotechnical Engineering
- Built Environment and Construction Engineering
- Physics
- Design and Technologies
- Urban Planning, Design, and Policy

Andrea De Toni · Andrea Arcidiacono ·
Silvia Ronchi
Editors

Nature-Positive Cities: Adaptive Spatial Planning in Italy for an Ecological Urban Transition

Editors
Andrea De Toni
Department of Architecture and Urban Studies (DAStU)
Politecnico di Milano
Milan, Italy

Andrea Arcidiacono
Department of Architecture and Urban Studies (DAStU)
Politecnico di Milano
Milan, Italy

Silvia Ronchi
Department of Architecture and Urban Studies (DAStU)
Politecnico di Milano
Milan, Italy

ISSN 2191-530X ISSN 2191-5318 (electronic)
SpringerBriefs in Applied Sciences and Technology
ISSN 2282-2577 ISSN 2282-2585 (electronic)
PoliMI SpringerBriefs
ISBN 978-3-032-06616-9 ISBN 978-3-032-06617-6 (eBook)
https://doi.org/10.1007/978-3-032-06617-6

This work was supported by Politecnico di Milano.

© The Editor(s) (if applicable) and The Author(s) 2026. This book is an open access publication.

Open Access This book is licensed under the terms of the Creative Commons Attribution-NonCommercial-NoDerivatives 4.0 International License (http://creativecommons.org/licenses/by-nc-nd/4.0/), which permits any noncommercial use, sharing, distribution and reproduction in any medium or format, as long as you give appropriate credit to the original author(s) and the source, provide a link to the Creative Commons license and indicate if you modified the licensed material. You do not have permission under this license to share adapted material derived from this book or parts of it.

The images or other third party material in this book are included in the book's Creative Commons license, unless indicated otherwise in a credit line to the material. If material is not included in the book's Creative Commons license and your intended use is not permitted by statutory regulation or exceeds the permitted use, you will need to obtain permission directly from the copyright holder.

This work is subject to copyright. All commercial rights are reserved by the author(s), whether the whole or part of the material is concerned, specifically the rights of translation, reprinting, reuse of illustrations, recitation, broadcasting, reproduction on microfilms or in any other physical way, and transmission or information storage and retrieval, electronic adaptation, computer software, or by similar or dissimilar methodology now known or hereafter developed. Regarding these commercial rights a non-exclusive license has been granted to the publisher.

The use of general descriptive names, registered names, trademarks, service marks, etc. in this publication does not imply, even in the absence of a specific statement, that such names are exempt from the relevant protective laws and regulations and therefore free for general use.

The publisher, the authors and the editors are safe to assume that the advice and information in this book are believed to be true and accurate at the date of publication. Neither the publisher nor the authors or the editors give a warranty, expressed or implied, with respect to the material contained herein or for any errors or omissions that may have been made. The publisher remains neutral with regard to jurisdictional claims in published maps and institutional affiliations.

This Springer imprint is published by the registered company Springer Nature Switzerland AG
The registered company address is: Gewerbestrasse 11, 6330 Cham, Switzerland

If disposing of this product, please recycle the paper.

Foreword

This book offers an opportunity to acknowledge the importance of studies and positions that have addressed problems related to climate change and their accumulation in urban and territorial practices over several decades. The set of contributions not only provides a focused cross-section of ongoing work but also raises questions about how it was initiated, how it is conducted and how its problems are displayed.

More than 50 years have passed since the United Nations Conference on the Human Environment held in Stockholm in 1972. This was the same year that the Club of Rome published research bringing the world's attention to an issue with disturbing implications, presenting it with a blunt title: 'The Limits to Growth'. Acknowledging and preparing for this warning—at a time when the global economy was shaped by ways to obtain the energy necessary for development from fossil fuels and, especially, the idea of human control over nature was well established—proved to be a difficult prospect to accept and process. After many years, the consequences of global warming and its drastic effects on many parts of the world have become clear, it is 'normal' to refer to climate change in weather forecasts and public debate, but policies are struggling. Openly denialist positions have decreased, but they have given way to delaying tactics (moving target dates forward) and/or blind confidence in technological progress, implying the idea that a lot can change, but without affecting the model of development. The extraordinary undertaking that the current revolution entails, suggests that we carefully consider the difficulties being encountered.

One aspect that is worth reflecting on concerns the origin of this angle and the subjects who introduced it, due to the consequences it has had on the general approach.

As at other times in history, research marked the turning point. This research was conducted by a large group of experts from various disciplines that met as a 'club' in Rome in 1968, at the behest of an entrepreneurial economist and an expert in science and technology. The members then made use of computer processing carried out by researchers at the Massachusetts Institute of Technology. The commitment and personal contacts of the club members encouraged dissemination of the resulting knowledge and their convictions, finding a 'sounding board' in international bodies.

By their nature, such bodies initiate action in member countries to promote collaboration in the technical, financial and regulatory fields, relying on individual will and initiative. In the last 50 years, we have witnessed the percolation of issues from international bodies to both governmental institutions and to social practices and public debate along different winding paths in each, loading the issues with interpretations, segmenting and specifying them in different contexts. This leads to an extremely complex decision-making chain, both direct and indirect, that is worth focusing on to understand the way the great objective of reconciling cities with nature—an objective that changes the paradigm and 'disrupts' the given order—is translated.

The development of the principles enunciated in research, experimentation and proposals, which have led to a new branch of urban and territorial studies, must be attributed to the push by the European Union, whose bodies, starting with the Directorates-General, have translated the objectives outlined by the United Nations into programmes and projects. This has yielded a series of initiatives aimed mainly at local institutions, those closest to the problems and citizens, featuring incentives for collaboration among participating subjects and the exchange/integration of approaches, knowledge, techniques and tools belonging to different related disciplinary fields. This has led to the confirmation of a method organised around strategy/action pairs and complex procedures, along with the construction of a language superimposed on disciplinary traditions. An approach using the persuasive power of a semantic utopia associated with economic and financial support has produced a new generation of innovative 'informal' tools. Impacts with the existing structure have involved both a double track, with new instruments placed alongside institutionalised tools for territorial governance, and introduction and fusion, when similarities and convergences are recognised.

In reading the 12 chapters of this book, I note not only continuous references to the issues of the international apparatus but also believe it is possible to recognise two main ways we can learn from the European 'lesson'. The first consists in redefining the disciplinary field, broadening and deepening it to include topics compatible with those already present in urban and territorial planning. The second consists of pursuing a lasting status for new instruments. The first includes the adoption of biodiversity, ecological connection and ecosystem services as guiding concepts in pointing landscape and material welfare in a new direction, while the second includes investments in Green Plans and Green and Blue Infrastructures. In my opinion, selecting aspects that can be dealt with in one's own field, together with reinterpreting and bending those already present, seem to be ways that urban and spatial planning has adopted to meet the challenge of climate change. A holistic approach was subjected to a process of reduction involving a selection of variables, assuming targets for politically significant timeframes, and standardisation.

I believe that all of this helps to explain the centrality of 'greenery', which is inextricably linked water and soil, in that they recall and give a complete meaning to the heterogeneous and dynamic set of design solutions defined as 'nature-based'.

There are now numerous case studies that enable us to build grids to synthesise the multifunctional nature of greenery and its ability to mitigate, rather than just adapt to the effects of climate change. We are used to using the terms 'adaptation' and 'mitigation' together, but the differences are not trivial. It is not just a question of defending ourselves to reduce the damage to people and property, but rather intervening in dangerous dynamics by understanding them.

Manuals prepared following reports and repertoires of good practices have permeated projects for transforming public spaces and, slowly, even the most difficult ones such as ground-level parking lots, where depaving, rebuilding the humus and planting change the performance and perception of the open space and its surroundings. These are achievements that the pandemic undoubtedly encouraged, albeit in a limited manner and not always professionally, but sufficient nevertheless to be convincing. Trends that are 'decorative' (green is evoked to validate any type of planned intervention), fashionable or ideological must not cause us to lose sight of the importance of a serious urban forestry policy, nor must it excessively abuse yet another semantic utopia as 'biocity': biocity is just one way to strategically denote urban and territorial regeneration.

Nevertheless, a critical attitude is indispensable in observing, for example, that the distribution of greenery is unequal, whether across parts of a city, across cities or across territories; that greenery also modifies urban values and its distribution must be considered with a view to equity and priority when resources are scarce; that the necessary massive operations call for equally massive private investments; that resistance to extreme events is possible with suitable species of 'healthy' (i.e. maintained) plants of the right age planted in appropriate locations; that flora is associated with fauna and that the metabolic dimension should be a cornerstone of the biocity; that urban forestry and densification are not two sides of the same coin; etc.

The European method—strategies/actions/targets/assessment/monitoring—and the introduction of thematic programmes and projects aimed directly at local institutions have encouraged the proliferation of informal tools and the development of new skills, but they have fuelled the idea that this will solve the problem of the inadequacy of traditional forms of territorial governance. A gap has formed, more or less consciously, between the search for other tools and a new form of the plan. However, the primary objective of limiting atmospheric warming by reducing the emission of greenhouse gases affects the economy and social well-being in a web that can only be considered by working with the tools held by the bodies responsible for territorial governance.

This convergence does not seem difficult with regard to municipal urban planning: the forerunners of the current green plans were already settled in the general urban plans. It is more problematic, however, to 'find a home' for green and blue infrastructure, which are indispensable for connectivity and biodiversity, as well as potential vehicles for regenerating other infrastructure under and above ground and the Italian landscape in general. They could support and connect the physical structures that allow environmental, economic and social systems to function, the framework of a country whose lithological and geomorphological features underlie serious instabilities. We have many specialised maps and national and regional sectors have accumulated knowledge related to their related fields, but nature knows no bounds and interdependence dominates. This is perhaps the most difficult issue to tackle: governing green and blue infrastructures.

Urban planners began to deal specifically with the relationship between plan contents and administrative boundaries with the great settlement expansion, but they seem to have grown timid with their unsuccessful experience with territorial districts. Faced with the disappointing results of the Metropolitan cities and the ineffective role of the Provinces in planning, the Regions—whose boundaries roughly align with historical borders and are not entirely detached from geographical areas—may hold decisive importance. Only by fully recovering the territorial dimension in regional strategies and resuming the country-wide experience of 'territorial projects' through a critical lens can we obtain any answers. In territorial, regional and interregional projects, urban and spatial planning should include environmental policies among their features. The Nature Restoration Law approved by the European Council in 2024, the first request for national governments to prepare plans expressed in terms of law, could/should offer a more solid foundation for regional initiatives, helping to overcome sector divisions and establish a solid link between landscape plans and territorial strategies.

Patrizia Gabellini
Honorary Professor of Urbanism
Politecnico di Milano
Milan, Italy

Acknowledgments

This book has received funding from the Project 'National Biodiversity Future Center—NBFC' funded under the National Recovery and Resilience Plan (NRRP), Mission 4 Component 2 Investment 1.4, Spoke 5—Call for Tender No. 3138 of 16 December 2021, rectified by Decree n.3175 of 18 December 2021 of Italian Ministry of University and Research funded by the European Union—NextGenerationEU; Project code CN_00000033, Concession Decree No. 1034 of 17 June 2022 adopted by the Italian Ministry of University and Research, CUP, D43C22001250001.

Contents

Aligning Urban Greening Policies with the EU Nature Restoration Regulation: Gaps and Prospects in Italy 1
Maria Chiara Pastore, Annarita Lapenna, and Luca Lazzarini

The Influence of the Urban Environment on Biodiversity: From a Systematic Literature Review to Spatial Planning Integration in Italy 13
Andrea De Toni, Chevonne Reynolds, Alessandro Alì, and Dan Chamberlain

Nexus Between Ecosystem Services Provision and Socio-Economic Variables: A Pathway for Equitable Planning 25
Silvia Ronchi and Marta Dell'Ovo

Living in Harmony with Nature? Climate, Biodiversity and Planning Futures 37
Fabiano Lemes de Oliveira

Ecological Connectivity Guides Spatial Planning 49
Alessandro Marucci and Lorena Fiorini

Planning Ecological Networks from Regional to Local Level: Reflections to Support Biodiversity for People and Nature 61
Serena D'Ambrogi and Anna Chiesura

Biodiversity and Landscape: Towards an Alliance in Italian Spatial Planning 77
Benedetta Giudice and Angioletta Voghera

Cities Walk, Forests Run: Trees and Forests as Nature-based Solutions in Transforming Biocities 89
Fabio Salbitano, Giuseppe Scarascia Mugnozza, and Marco Marchetti

Nature-Positive: Transforming Cities and Landscapes with Scalable Strategies and Projects. Insights from LAND's Case Studies 103
Andreas Otto Kipar, Valentina Galiulo, Gloria Signorini, and Daniele Galimberti

Role of Spatial Planning in Addressing Climate Challenges: A Study Concerning the Functional Urban Area of Cagliari 115
Sabrina Lai and Corrado Zoppi

Planning for Biodiversity: Strategies and Actions for Enhancing Nature in the Urban Plan of Varese (Italy) 127
Andrea Arcidiacono, Laura Pogliani, Silvia Ronchi, Stefano Salata, Andrea Benedini, Federico Ghirardelli, and Beatrice Mosso

Ecological Planning Strategies and Nature-based Solutions in the Context of Climate Change Resilience 139
Davide Geneletti, Chiara Cortinovis, Chiara Parretta, Simone Caridi, Giuseppe Formetta, Lorenzo Giovannini, Lia Laporta, Alfonso Vitti, and Jarumi Kato-Huerta

About the Editors

Andrea De Toni (Ph.D., *cum laude*) is an Assistant Professor in Urban Planning, Department of Architecture and Urban Studies (DAStU), Politecnico di Milano, former Pro-Rector's Delegate for Sustainability (Territorial Pole of Mantua). Involved in several prestigious national and international research projects, and part of the winning team of the British Ecological Society's Synthesis Grant, she sees interdisciplinarity as the innovative core of her research. For years, she has been working closely with ecologists and foresters, integrating biodiversity and ecological knowledge into urban and regional planning. Much of her work has been based upon the analysis of quantitative and qualitative socio-economic and environmental geo-referenced data on different scales and the engagement of relevant stakeholders to support the decision-making processes in selecting planning priorities.

Andrea Arcidiacono (Ph.D.) is a Full Professor in Urban Planning, at the Department of Architecture and Urban Studies (DAStU), Politecnico of Milano; a Coordinator of the Bachelor of Science programme in 'Urban Planning'; AUIC School Member of the Board of the Ph.D. Programme in Urban Planning, Design and Policy UPDP; and the Director of the LabPPTE (Landscape Plans Territories Ecosystems Lab) and Land Take Research Centre CRCS. From 2016 to 2023, he has been Vice-President of the National Institute of Urban Planning, INU. He is a Member of the Scientific Editorial board of the journal *Urbanistica*. His research interests include urban planning and design, landscape and environmental planning, green infrastructure design, ecosystem services analysis and nature-based solutions for spatial planning, policies and planning strategies for land take limitation. He is involved in European and national research programmes and has authored over 160 scientific publications.

Silvia Ronchi (Ph.D.) is an Assistant Professor in Urban Planning at the Department of Architecture and Urban Studies (DAStU), Politecnico di Milano. She collaborates on and leads several national and international research projects as a scientific coordinator. In 2017, she earned a Ph.D. in Urban Planning, Design and Policy from the Politecnico di Milano, with a dissertation on the integration of ecosystem services into spatial planning. From 2015 to 2018, she collaborated with the Joint

Research Centre of the European Commission. In January 2016, she was a visiting researcher at the University of Salzburg (Austria). She is a Member of the Scientific Editorial Board of the journal *Territorio* (FrancoAngeli). Her research interests focus on ecosystem services, environmental assessment, landscape and urban planning, as well as soil sealing and land take processes. She has (co-)authored over 110 scientific publications.

The Editors, Andrea De Toni, Andrea Arcidiacono and Silvia Ronchi, are part of the National Biodiversity Future Center (NBFC), Spoke 5, Biodiversity and Human Well-Being.

Aligning Urban Greening Policies with the EU Nature Restoration Regulation: Gaps and Prospects in Italy

Maria Chiara Pastore, Annarita Lapenna, and Luca Lazzarini

Keywords Urban biodiversity · Urban greening plans · Urban greening policies · Green urban spaces · Nature restoration regulation

1 Introduction: The Challenge of Nature Restoration in the EU

There is growing recognition that the interconnected biodiversity and climate crises demand international collaboration and effective multilevel governance. This places supranational institutions such as the European Union in a pivotal position to drive national efforts by establishing robust policies and binding targets aimed at addressing the regional and local impacts of these global challenges [1].

In 2019, building on previous programmes and actions launched by the European Commission to address climate change, the EU Green Deal formalized a new phase of European policies by placing climate and environmental issues at the center, with the aim of influencing public policies at national and sub-regional levels. By developing a complex system, the EU Green Deal outlines eight closely interconnected

---

M. C. Pastore · A. Lapenna (✉)
Department of Architecture and Urban Studies (DAStU), Politecnico di Milano, National Biodiversity Future Center (NBFC), Florence, Italy
e-mail: annarita.lapenna@polimi.it

M. C. Pastore
e-mail: mariachiara.pastore@polimi.it

L. Lazzarini
Laboratorio Di Simulazione Urbana 'Fausto Curti', Department of Architecture and Urban Studies (DAStU), Politecnico di Milano, National Biodiversity Future Center (NBFC), Florence, Italy
e-mail: luca.lazzarini@polimi.it

© The Author(s) 2026
A. De Toni et al. (eds.), *Nature-Positive Cities: Adaptive Spatial Planning in Italy for an Ecological Urban Transition*,
PoliMI SpringerBriefs, https://doi.org/10.1007/978-3-032-06617-6_1

areas of intervention: Climate, Energy, Environment and Oceans, Agriculture, Transport, Industry, Research and Innovation, and Financing and Regional Development. Among these, the issue of urban biodiversity is notably addressed in the "Environment and Oceans" sector through the EU Biodiversity Strategy for 2030 [2]. As part of its effort to halt biodiversity loss and tackle habitat degradation, the EU Biodiversity Strategy (EU BS) outlines a comprehensive set of commitments and actions aimed at putting Europe's biodiversity on a path to recovery by 2030 [3]. The Strategy is structured around four key pillars—Protect Nature, Restore Nature, Enable Transformative Change, and EU Action to Support Biodiversity Globally—each supported by specific commitments and objectives. These strategies are accompanied by explanations of the necessary financial resources and the actions required to achieve them. To monitor the over 100 actions and 17 targets identified in the Strategy, two main online progress tools (the *actions tracker* and the *dashboard*) have been developed, regularly reporting on the Strategy progress to EU institutions and national member states. While the *actions tracker* is designed to monitor the implementation of the actions with information on the status and year of implementation, the *dashboard* has the objective to monitor the progress of the 17 targets through a set of indicators. As highlighted by Marei Viti et al. [4], while the actions tracker is a mature tool, the dashboard has currently many indicators that had not been specifically developed to monitor the Strategy and scientific input is needed to ensure policy tracking and transparent and data-driven monitoring of the targets. Indeed, an effective system of targets and indicators alone is not sufficient to ensure that the EU BS is implemented but a collective and collaborative effort is also essential. Hermoso et al. [5] (p. 268), pointed out that the success of the EU Biodiversity Strategy in achieving its ambitious objectives depends on the "capacity of the EU Member States to plan strategically the implementation of conservation measures under limited and uncertain budgets, better engage with the general public, and avoid or solve potential conflicts with other socio-economic objectives and different sectoral policies".

One of the crucial topics underlying the Strategy is restoration, for which a proposal for legally binding nature restoration targets is included, as an opportunity to increase EU resilience and contribute to climate change mitigation and adaptation. After years of discussion and intense negotiations, in June 2024 the European Council has finally adopted the Nature Restoration Regulation (NRR). The Regulation introduces binding restoration targets for specific habitats and species through measures that should cover at least the 20% of the EU land and sea areas by 2030, and ultimately all ecosystems in need of restoration by 2050. Implementation, a crucial aspect of the regulation, is carried out at the national level through dedicated National Restoration Plans (NRP). Member states are required to submit these plans to the European Commission by mid-2026, demonstrating how they intend to meet the established targets. The NRR defines restoration as "the process of actively or passively assisting the recovery of an ecosystem in order to improve its structure and functions, with the aim of conserving or enhancing biodiversity and ecosystem resilience, through improving an area of a habitat type to good condition, re-establishing favourable reference area, and improving a habitat of a species to sufficient quality and quantity" [6] (p. 19). Although terms such as quality, quantity, and favourable reference area are defined and explained in the articles of the Regulation, some authors argue that the NRR adopts a rather broad interpretation of the concept of restoration, leaving too much room for interpretation, which could lead to inconsistent implementation across countries or legal uncertainty for national and local institutions [7]. Furthermore, they contend that the legislation does not fundamentally alter the EU's existing approach to nature. Instead, it largely reaffirms obligations already set out in previous legislation and allows for several exceptions and derogations, which may limit its overall transformative potential. These provisions reduce the obligations of member states in specific situations, such as those involving national defence, security, or infrastructure projects aimed at accelerating the deployment of renewable energy, which could otherwise hinder nature restoration efforts. According to Penca and Tănăsescu [7], despite the difficulty of the Regulation in addressing the root causes of nature's decline beyond the promotion of restoration actions, some aspects of the NRR are potentially far reaching in stimulating member countries to debate around the factors and dynamics which bring to biodiversity loss and ecosystem degradation, complementing existing Nature Directives (Birds Directive, 1979 and Habitats Directive, 1992) with a broader and more proactive approach. Within this complex framework, this chapter aims to present and discuss the most significant innovations introduced by the NRR concerning urban ecosystems. In particular, it focuses on a specific instrument—Urban Greening Plans, recently renamed "Urban Nature Plans" by the European Commission—as a strategic planning tool that could support the implementation of the objectives and targets set out by the NRR in relation to urban environments.

The chapter is articulated in four sections. Section 2 introduces urban ecosystems as one of the key geographical scopes of the NRR by focusing on two crucial challenges associated with regulation's implementation. Section 3 investigates the Urban Greening Plans as overarching strategic frameworks introduced by the EU Biodiversity Strategy for guiding the transition toward nature-positive cities. It also analyses the recent transition from Urban Greening Plans to Urban Nature Plans promoted by the European Commission and some open issues concerning the capacity of these instruments to contribute to the implementation of the NRR. In Sect. 4 we examine the Italian context by focusing on the *Piani del Verde* and their potential to play a role in NLR implementation at the national level. The chapter ends with a reflection on the opportunity to interpret *Piani del Verde* as platforms for overcoming the current fragmentation of responsibilities and fostering more synergistic and cross-sectoral governance arrangements in the field of nature preservation in cities and urban areas.

2 Urban Ecosystems: A Geographical Scope of the Nature Restoration Regulation

The NRR establishes binding rules for the long-term recovery of biodiversity and the resilience of ecosystems across terrestrial and marine areas, through the restoration of degraded ecosystems (Article 1, NRR). Among the spatial domains targeted by the regulation, urban ecosystems are explicitly identified. The increasing recognition of urban greenspaces as critical for biodiversity conservation [8] calls for the development of targeted Regulation for these urban ecological systems.

Research in this field has expanded by distinguishing three paradigms: the ecology *in*, *of*, and *for* the city [9]. *Ecology in the city* focuses on the study of green and aquatic patches within urban contexts. *Ecology of the city* views the entire urban mosaic as a socio–ecological system, integrating biotic, social, and built components. Finally, *ecology for the city* emphasizes the co-production of knowledge and the collaborative definition of goals and strategies that support sustainability and resilience.

The approach advocated by the NRR aligns with the paradigm of "ecology for the city" [10], which does not merely describe ecological processes in cities, but encourages active engagement by planners, citizens, and policymakers in co-designing sustainable solutions. The NRR recognizes the need for restoration in urbanized contexts, aiming to enhance biodiversity and ecological resilience within the urban fabric. Cities are thus both sources of ecological pressure and potential sites for restoration.

The Article 8 stipulates that, by the end of 2030, Member States must ensure no net loss in the total national area of urban green space or tree canopy cover within urban ecosystem zones. From 2031 onwards, these areas must show a positive trend in coverage.

These requirements raise two key challenges. The first concerns the definition of degraded urban ecosystems and the criteria for identifying restoration areas. As of now, there is no unified regulation providing a comprehensive and operational definition of "degraded ecosystem". The concept is interpreted through the Habitats Directive (92/43/EEC) [11], which includes indicators of "unfavourable" conservation status, and the EU Biodiversity Strategy for 2030, which refers to degraded ecosystems as those unable to deliver essential services or meet ecological integrity standards. In urban contexts, degradation is not limited to ecological impoverishment but also refers to the loss of regulatory, cultural, or social functions [12].

The second challenge lies in delineating urban ecosystem zones. By 2026, Member States are required to map and designate these zones across all their cities, towns, and suburbs. However, the NRR does not provide standardized criteria for what constitutes an urban ecosystem. This opens a critical implementation risk: national authorities may interpret urban ecosystems as broad-scale entities encompassing both dense urban cores and adjacent low-density municipalities. While such aggregations may align with the functional governance structures (e.g., metropolitan commuting zones), they risk fulfilling restoration targets without significantly improving the most ecologically compromised urban areas—thus undermining the transformative intent of the regulation.

Within the NRR implementation framework, the role of Member States—already emphasized in the EU Biodiversity Strategy for 2030—is fundamental in identifying which parts of urban areas qualify as ecosystems in need of restoration. This involves the integration of urban planning tools, green infrastructure strategies, and mapping techniques. Urban ecosystems are complex systems for two main reasons: first, they encompass a significant variety of specific natural ecosystem types; second, they are socio-ecological systems in which the cultural and social dimensions play a decisive role in each local context [13]. For this reason, observation and actions at the municipal and local scales are crucial for the success of restoration planning.

In light of these reflections, Urban Greening Plans—recently re-named Urban Nature Plans by the European Commission and already adopted or under development in various European cities—can serve as valuable platforms for consolidating knowledge on urban ecosystems, supporting political decision-making, and providing technical-scientific arguments for national restoration planning and subsequent implementation and monitoring phases.

3 Urban Greening Plans as Instruments for Guiding the Transition Toward Nature-Positive Cities

In the first section, we have mentioned the crucial role that the target identification and the progress tools have in monitoring the implementation of the EU BS. One important goal of the Strategy is the greening of urban and periurban areas for which the systematic integration of healthy ecosystems, green infrastructure and nature-based solutions into all forms of urban planning is promoted.

To achieve this, Target 14 of the Strategy highlights that "cities with at least 20,000 inhabitants should adopt ambitious greening plans". Urban greening plans (UGP) are defined as "overarching frameworks articulating, formalizing, and showcasing a city's commitment to promoting and protecting biodiversity and urban greening" with the aim to ensure that "urban planning processes systematically incorporate and promote green infrastructure thinking and nature-based solutions" [3]. While the term "ambitious" has sparked debate among policymakers and experts [14–17] the Commission has identified three main actions to achieve this target: (i) commissioning technical Guidance to assist cities in developing ambitious UGPs; (ii) creating an online urban greening platform to provide cities with information and support for setting up their plans; and (iii) developing the UGPs themselves.

Looking more in detail at the Guidance, the interesting point is the identification of 10 steps for establishing an UGP (see Fig. 1) which cover, respectively, the phases of preparation, action planning, implementation and monitoring. These three phases are united by a co-creation approach where the engagement of different key stakeholders

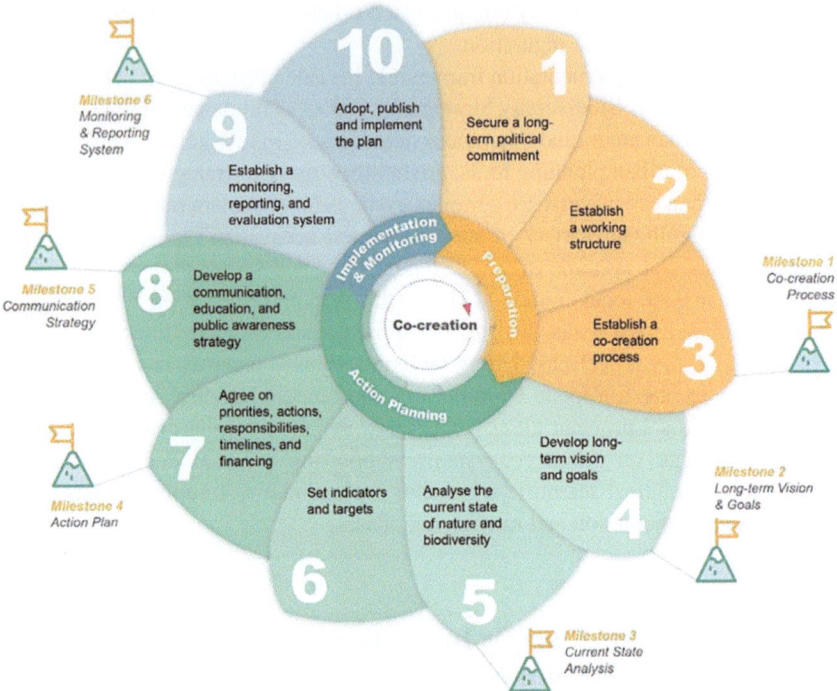

Fig. 1 Concept diagram with the steps needed to prepare and implement an urban nature plan. *Source* European Commission [19]

within a broad participatory process is highlighted to ensure that diverse perspectives, knowledge, and experiences are positively integrated into the urban greening process.

Following the publication of a draft version of the Guidance by the European Commission in 2021—developed in collaboration with Eurocities and ICLEI—, a new version was released in April 2025 on the Urban Nature Platform. Notably, the title "Urban Greening Plans" has been replaced by "Urban Nature Plans". Aside from this change and a few minor revisions in the document, the structure, content, and objectives of the Guidance have remained largely unchanged. One crucial point emerging from a discourse analysis of the Guidance concerns its thematic focus which remains notably centered on the preservation and enhancement of urban green spaces. Indeed, its scarce capacity to address the complex dynamics of nature in relation to human activities is underlined. This is particularly evident in the indicators cited, which include the percentage of urban green space in the city, tree canopy cover within the municipality, the number of newly planted trees, and the extent of protected natural areas on public land. Notably absent is any consideration of the root causes of nature's decline in cities, as well as of the patterns of human–nature relationships and their capacity to guide or catalyze practices of care and stewardship which tackle the root causes already mentioned [7, 13]. Hence, a crucial point we want to highlight is the need for Urban Nature Plans to move beyond a narrow focus on green spaces, toward a more robust acknowledgment of the objectives and strategies related to the plant and animal diversity that characterizes each town or city. Nature encompasses far more than just green areas, and addressing it effectively requires a comprehensive set of strategies, indicators, and targets that can confront biodiversity loss and enhance the capacity of urban ecosystems to support diverse non-human species, fostering multispecies coexistence [18].

Another limitation of the Guidance lies in its lack of incorporation of a restoration-based approach and, more broadly, its insufficient recognition of the Urban Nature Plan as a potential implementation tool for the NRR. This connection is mentioned only briefly in the section on core targets, where it is stated that "[a]s a minimum, targets should be set in line with national and European environmental policies and regulations relating to nature and biodiversity. This means aligning the Urban Nature Plan with the upcoming NRR targets on 'urban ecosystems' and measuring and monitoring the core criteria set out for nature and biodiversity in the Green City Accord" [19] (p. 11). While clear guidance has proven necessary to clarify the targets related to urban ecosystems outlined in the Regulation and translate them to the subnational and municipal levels, the responsibility for doing so has ultimately been left to the NRP of the Member States, where guidance on how to implement the targets of the NRR at the local level is expected to be included.

4 The Italian *Piani del Verde* and the National Restoration Plan

In Italy, *Piani del Verde* (PdVs) are municipal thematic planning tools aimed at the integrated management of urban and peri-urban green spaces. Although developed on a voluntary basis, their role was first formally outlined in the *Guidelines for the Management of Urban Green Areas* [20], which define minimum contents and strategic objectives for sustainable green infrastructure planning and governance. Currently adopted by a limited number of municipalities [21], PdVs operate within a fragmented policy framework where urban biodiversity is often addressed indirectly [2]. However, the increasing attention to nature-urban relationship in European policy—notably in the EU Biodiversity Strategy for 2030 and the NRR—has begun to reframe these plans as potential strategic instruments to guide the local governments' action toward the preservation and enhancement of urban biodiversity.

Despite their sectoral and soft nature [13], PdVs can contribute to transformative changes in urban socio-ecological systems and governance arrangements. As anticipated in the *Urban Nature Plan Guidance*, PdVs may function as experimental platforms for integrating biodiversity restoration objectives at the local scale, by collecting baseline data, identifying spatial priorities, and outlining context-specific interventions.

In Italy, the process of drafting the National Restoration Plan, as required by the NRR, is currently in its early stages. In May 2025, the European Commission published an *Implementing Regulation* (EU) 2025/912 [22], establishing a standardized format for the submission of NRP by Member States. The Regulation sets out a structured template, which includes: (i) a synthesis of the national context; (ii) mapping and assessment of degraded ecosystems; (iii) target-setting at both national and ecosystem levels; (iv) identification of restoration measures; (v) mechanisms for governance, financing, and monitoring; and (vi) alignment with existing EU legislation. In this context, Article 8 of the NRR—which focuses specifically on urban ecosystems—calls for targeted restoration measures to be implemented in urban and peri-urban areas to improve biodiversity, ecological connectivity, and access to green infrastructures.

Although Italian municipalities have not yet been formally involved in the NRP drafting process, the PdVs already in place can support this phase by providing a consistent knowledge base and operational framework. Many of these plans include data on the extent, typologies, and ecological status of urban green areas, as well as participatory processes and spatial analyses that align with the requirements of the EU format. As such, PdVs could act as foundational inputs for mapping, priority-setting, and planning of locally appropriate restoration actions, in line with Article 8 of the NRR.

Moreover, PdVs could also play a critical role during the implementation and monitoring phases of the NRP. Their embeddedness in local governance structures makes them suitable tools to support multi-level coordination, track progress towards restoration targets, and ensure the consistency of actions across different sectors. For example, the *Piano del Verde* of Brescia [23] incorporates performance indicators, ecological criteria, and interdepartmental governance arrangements that could serve as a model for integrating local green infrastructure plans into national-scale restoration frameworks.

Nonetheless, this requires formal recognition of the role of the PdVs in national policy frameworks. The absence of a national regulatory framework supporting their systematic adoption, combined with the limited capacity of many local governments to implement them, remains a major barrier. Integrating local PdV into the enabling tools of the NRP could act as a strategic lever to enhance their uptake and strengthen municipal contributions to EU targets. This would also foster a more place-based implementation of the NRR, aligning it with the territorial and socio-ecological specificities of Italian cities.

5 Conclusions

In light of the transitional phase currently characterizing the development of the National Restoration Plan, PdVs can assume a strategic role as platforms for the integration of multilevel policies and local actions. Despite their still voluntary and experimental nature, the PdVs offer a concrete opportunity to incorporate objectives, targets, and obligations stemming from the new European regulatory framework on ecosystem restoration, with particular reference to urban contexts.

From this perspective, the PdV could be reformed not only to strengthen its effectiveness as a foundational and guiding tool within the framework of statutory urban planning, but also to ensure greater consistency with the principles and aims of the NRR and the EU Biodiversity Strategy for 2030. Such reform would promote the integration of planning and environmental instruments, overcoming the current fragmentation of responsibilities and fostering more synergistic, cross-sectoral and nature-positive governance arrangements [24].

A closer alignment between the PdV and the NRP would enable the convergence of strategies and tools that currently operate at different scales—national and municipal—but which share a common orientation towards the conservation and enhancement of the ecological quality of natural and semi-natural ecosystems. This would require a revision of the minimum contents of the PdV to include ecological status indicators, restoration scenarios, and monitoring systems with targets aligned with those envisaged by the NRP. In this perspective, the PdV could evolve into a hybrid instrument: simultaneously strategic and operational, capable of capturing local specificities while also responding to the supranational commitments undertaken by the National government. A potential improvement of the PdV thus represents a crucial step toward consolidating integrated urban ecological planning, aimed not only at managing existing green infrastructure but also at restoring degraded urban ecosystems—thereby actively and structurally contributing to the achievement of the European environmental restoration goals.

Acknowledgements This research has received funding from the Project "National Biodiversity Future Center—NBFC" funded under the National Recovery and Resilience Plan (NRRP), Mission 4 Component 2 Investment 1.4—Call for tender No. 3138 of 16 December 2021, rectified by Decree n.3175 of 18 December 2021 of Italian Ministry of University and Research funded by the European Union—NextGenerationEU; Project code CN_00000033, Concession Decree No. 1034 of 17 June 2022 adopted by the Italian Ministry of University and Research, CUP, D43C22001250001.

Authorship Contribution Statement Conceptualisation, M.C.P., A.L., L.L.; methodology, M.C.P., A.L., L.L.; validation, M.C.P.; investigation, M.C.P., A.L., L.L.; resources, M.C.P.; data curation, A.L., L.L.; writing—original draft preparation, A.L., L.L.; writing—review and editing, M.C.P.; supervision, M.C.P.; project administration, M.C.P.; funding acquisition, M.C.P. All authors have read and agreed to the published version of the manuscript.

References

1. Ferraro G, Failler P (2024) Biodiversity, multi-level governance, and policy implementation in Europe: a comparative analysis at the subnational level. J Publ Policy 44(3):546–572. https://doi.org/10.1017/S0143814X24000072
2. Lapenna A (2025) Politiche pubbliche per la biodiversità urbana: una storia recente. In: Pastore MC, Lapenna A, Lazzarini L, Mahmoud IH, Zanotto F (eds) Città biodiverse: politiche, piani, progetti e processi di co-creazione. Mimesis, Milano

3. European Commission (2021) EU biodiversity strategy for 2030. Bringing nature back in our lives. Publications Office of the European Union, Luxembourg
4. Marei Viti M, Gkimtsas G, Liquete C et al (2024) Introducing the progress monitoring tools of the EU Biodiversity Strategy for 2030. Ecol Ind 164:112147. https://doi.org/10.1016/j.ecolind.2024.112147
5. Hermoso V, Carvalho SB, Giakoumi S et al (2022) The EU biodiversity strategy for 2030: opportunities and challenges on the path towards biodiversity recovery. Environ Sci Policy 127:263–271. https://doi.org/10.1016/j.envsci.2021.10.028
6. European Parliament and Council (2024) Regulation (EU) 2024/1991 of the European Parliament and of the Council of 24 June 2024 on nature restoration and amending Regulation (EU) 2022/869. Office Journal of the European Union, Bruxelles
7. Penca J, Tănăsescu M (2025) The transformative potential of the EU's nature restoration law. Sustain Sci 20:643–647. https://doi.org/10.1007/s11625-024-01610-6
8. Klaus VH, Kiehl K (2021) A conceptual framework for urban ecological restoration and rehabilitation. Basic Appl Ecol 52:82–94. https://doi.org/10.1016/j.baae.2021.02.010
9. Pickett STA, Cadenasso ML, Childers DL, McDonnell MJ, Zhou W (2016) Evolution and future of urban ecological science: ecology in, of, and for the city. Ecosyst Health Sustain. 2(7):e01229. https://doi.org/10.1002/ehs2.1229
10. Childers DL, Cadenasso ML, Grove JM, Marshall V, Mcgrath B, Pickett STA (2015) An ecology for cities: a transformational nexus of design and ecology to advance climate change resilience and urban sustainability. Sustainability 7:3774–3791
11. European Parliament and Council (1992) Council directive 92/43/EEC of 21 May 1992 on the conservation of natural habitats and of wild fauna and flora. Official Journal L 206, 22.07.1992, pp 7–50
12. McPhearson T, Haase D, Kabisch N, Gren Å (2016) Advancing understanding of the complex nature of urban systems: integrating ecosystem services, complexity theory, and social–ecological systems. Ecol Ind 70:566–573. https://doi.org/10.1016/j.ecolind.2016.02.054
13. Pastore MC, Lapenna A, Lazzarini L (2024) The green ambition. Il contributo dei Piani del verde alla biodiversità urbana in Italia. In: Pisano C, De Luca G (eds) Progettare nel disordine - Progettare il disordine. Riordinare le fragilità urbane. INU Edizioni, Roma
14. German Environment Agency (2021) Tackling the climate and biodiversity crises in Europe through Urban Greening Plans. Position Paper, Dessau: Umweltbundesamt
15. Bellè BM, Deserti A (2024) Urban greening plans: a potential device towards a sustainable and co-produced future. Sustainability 16:5033. https://doi.org/10.3390/su16125033
16. Chiesura A et al (2024) I Piani comunali del verde: strumenti per riportare la natura nella nostra vita? Quaderno ISPRA 33/2024
17. Mahmoud I, Dubois G, Liquete C et al (2025) Understanding collaborative governance of biodiversity-inclusive urban planning: methodological approach and benchmarking results for urban nature plans in 10 European cities. Urban Ecosyst. 28:17. https://doi.org/10.1007/s11252-024-01656-5
18. Pastore MC, Lapenna A, Lazzarini L, Mahmoud I, Zanotto F (eds) (2025) Città biodiverse. Politiche, piani, progetti e processi di co-creazione. Mimesis, Milano
19. European Commission (2024) Urban nature plans—guidance for cities to help prepare an urban nature plan. Publications Office of the European Union, Bruxelles. https://data.europa.eu/doi/10.2779/353044
20. Comitato per lo sviluppo del verde pubblico, MATTM (2017) Linee guida per la gestione del verde urbano e prime indicazioni per una pianificazione sostenibile
21. Pastore MC, Lapenna A, Lazzarini L (2025) I Piani del Verde in Italia: tra assetto strategico e normativo. In: AA.VV. (eds) Il verde nella città che cambia. Képos Libro Bianco del Verde, Roma
22. European Commission (2025) Commission implementing regulation (EU) 2025/912 of 19 May 2025 laying down rules for the application of Regulation (EU) 2024/1991 of the European Parliament and of the Council as regards a uniform format for the national restoration plan. https://eur-lex.europa.eu/legal-content/EN/TXT/?uri=CELEX%3A32025R0912&qid=1747725439193

23. Comune di Brescia (2025) Piano del Verde e della Biodiversità. https://www.comune.brescia.it/aree-tematiche/verde-e-parchi/piano-del-verde-e-della-biodiversita#documenti
24. Frantzeskaki N, Hölscher K, Wittmayer JM, Avelino F, Bach M (2018) Transition management in and for cities: introducing a new governance approach to address urban challenges. In: Frantzeskaki N, Hölscher K, Bach M, Avelino F (eds) Co-creating sustainable urban futures. Future city, vol 11. Springer, Cham. https://doi.org/10.1007/978-3-319-69273-9_1

Open Access This chapter is licensed under the terms of the Creative Commons Attribution-NonCommercial-NoDerivatives 4.0 International License (http://creativecommons.org/licenses/by-nc-nd/4.0/), which permits any noncommercial use, sharing, distribution and reproduction in any medium or format, as long as you give appropriate credit to the original author(s) and the source, provide a link to the Creative Commons license and indicate if you modified the licensed material. You do not have permission under this license to share adapted material derived from this chapter or parts of it.

The images or other third party material in this chapter are included in the chapter's Creative Commons license, unless indicated otherwise in a credit line to the material. If material is not included in the chapter's Creative Commons license and your intended use is not permitted by statutory regulation or exceeds the permitted use, you will need to obtain permission directly from the copyright holder.

The Influence of the Urban Environment on Biodiversity: From a Systematic Literature Review to Spatial Planning Integration in Italy

Andrea De Toni, Chevonne Reynolds, Alessandro Alì, and Dan Chamberlain

Keywords Land use planning · Resilience planning · Urban ecology · Ecosystem services · Green infrastructure · PGT

1 Introduction

The rapid expansion of urban areas worldwide, projected to triple by 2030, represents one of the most significant drivers of contemporary biodiversity loss. Driven by population growth, economic development, and infrastructural needs, urbanization transforms natural ecosystems into built environments at an unprecedented scale and pace

---

A. De Toni (✉) · A. Alì
Department of Architecture and Urban Studies (DAStU), Politecnico di Milano, 20133 Milano, Italy
e-mail: andrea.detoni@polimi.it

A. Alì
e-mail: alessandroali@ubistudio.it

A. De Toni · D. Chamberlain
NBFC—National Biodiversity Future Center, 90133 Palermo, Italy
e-mail: danieledward.chamberlain@unito.it

C. Reynolds
School of Animal, Plant and Environmental Sciences (APES), University of the Witwatersrand, Braamfontein 2001, South Africa
e-mail: chevonne.reynolds@wits.ac.za

D. Chamberlain
Department of Life Sciences and Systems Biology, Università di Torino, 10123 Turin, Italy

© The Author(s) 2026
A. De Toni et al. (eds.), *Nature-Positive Cities: Adaptive Spatial Planning in Italy for an Ecological Urban Transition*,
PoliMI SpringerBriefs, https://doi.org/10.1007/978-3-032-06617-6_2

[1–3]. This process fundamentally alters landscape structure through habitat fragmentation, loss of natural green spaces, and modification of ecological processes that support biodiversity [4–6]. Urban development creates novel environmental conditions characterised by altered microclimates through urban heat island effects, modified resource availability, and increased disturbance regimes that challenge existing species assemblages [5–7]. The resulting ecological pressures typically favour generalist species capable of adapting to urban conditions while disadvantaging specialist species with narrow habitat requirements, leading to biotic homogenization and generally reducing overall biodiversity in urban landscapes [2, 4, 5].

Recognition of cities' potential to support biodiversity has fundamentally shifted in both academic research and policy approaches to urban environments over recent decades. This shift has been driven by growing evidence that strategic urban planning can mitigate biodiversity loss while supporting essential ecosystem functions. While urbanization undeniably poses threats to native biodiversity, cities also create opportunities for biodiversity conservation through strategic planning, creation and management of urban green spaces and nature-based solutions [2, 8, 9]. Urban areas can provide critical habitat for certain species, support important ecosystem services such as pollination, and serve as stepping stones for wildlife movement across fragmented landscapes [8, 10]. Contemporary urban biodiversity research has evolved to examine not only the negative impacts of urbanization but also how urban planning and design can minimize biodiversity loss while maximizing conservation opportunities within urban environments [8, 9]. This growing understanding has informed policy initiatives worldwide, including the EU Biodiversity Strategy for 2030, with urban biodiversity conservation increasingly recognised as essential for achieving broader sustainability goals (e.g., UN SDG 11) and maintaining ecosystem services that support both urban populations and regional biodiversity networks.

The Member States of the European Union are therefore called upon to address the challenge of transposing directives and regulations, including the recent EU Nature Restoration Law. Among them, Italy is characterized on the one hand by intense anthropogenic pressure and high ecological fragmentation due to increasing land consumption [11], and on the other by a delay in integrating environmental objectives into strategic and municipal planning [12]. In particular, Lombardy Regional Law No. 12 of 2005 on soil sealing reduction requires an update that explicitly considers the protection of urban biodiversity. Significant knowledge gaps persist regarding the effective integration of biodiversity conservation measures into both existing planning frameworks and instruments, as reflected in the scientific literature and urban planning practice. Thus, the present study aims to identify the main threats to biodiversity in urban environments at fine scale through a systematic literature review, finally providing recommendations for integrating these insights into Italian spatial planning, specifically in relation to the Municipal Urban Plan (PGT).

2 Material and Method

A systematic literature review was performed by analysing data from studies conducted within the Italian context, to determine general trends on how urban characteristics impact urban biodiversity.

2.1 Search Protocol and Data Collection

The search for references was carried out on Scopus, using the string Search strings: "urban biodiversity" OR "biodiversity" AND "urban" OR "urban environm*" OR "urban feature*" OR "urban threat*" AND "impact*" AND "Italy". The PRISMA protocol was implemented for collecting data from scientific literature, basing the methodology on Zuniga-Palacios et al. [13] (see Fig. 1). The literature review was concluded in late May 2025.

In detail, concerning the (i) identification section: no publication conducting a literature review was identified with specific reference to the treatment of urban environmental threats to biodiversity in the Italian context. This study represents the first tentative review in this subject area; (ii) eligibility section: articles addressing topics related to marine and coastal contexts as well as volcanic risk were excluded

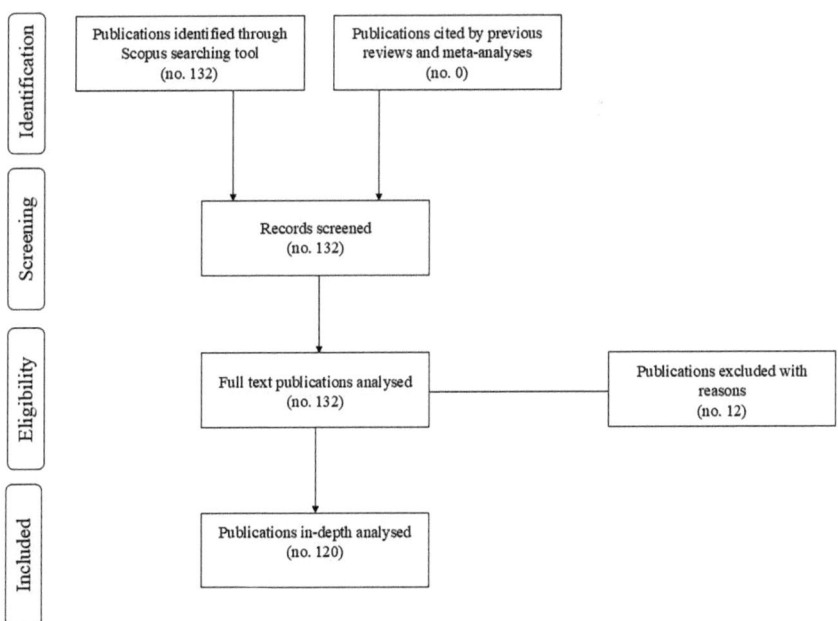

Fig. 1 PRISMA flowchart showing the search procedure. Authors' elaboration

from the analysis. The present review is a preliminary step toward discussing the integration of such threats into the PGT of the Lombardy Region, a region that does not border the sea and which does not host volcanoes.

3 Results

3.1 Urban Threats and Related Taxa Covered

Table 1 summarizes the main threats to urban biodiversity and the corresponding types of biodiversity affected. The threats to biodiversity have been grouped based on how they are addressed within the PGT. Indeed, given that many threats are interconnected (soil sealing includes not only general impermeabilization, but also the construction of infrastructure and urban sprawl), the classification was based on a combined analysis of threats and potential planning actions to be integrated into the PGT (Fig. 2).

The papers from the literature review were classified according to the type of response variable considered in relation to various threats arising from urbanization. There were nine categories in total: birds, invertebrates, mammals, plants, biodiversity general (either other taxa, or indicators of overall biodiversity), climate/air quality, land cover (usually measured through remote-sensed data), soils and 'other' (i.e., responses that did not fit into any of the other classifications). Of the 120 papers

Table 1 Summary of the main categories of threats and the corresponding types of biodiversity affected

Threat category	Biodiversity type affected
1. Threats related to soil sealing and loss of green areas, land use changes and urban sprawl	Soil biodiversity and ecosystem functioning; vegetation structure and landscape heterogeneity; pollinators and flower-associated insects; arthropods and invertebrate fauna
2. Microclimatic alterations, climate change impacts, urban pollution (air, water, soil) and artificial light at night	Urban biodiversity and ecological sensitivity; habitat diversity and ecosystem representation; pollinators and associated trophic interactions
3. Infrastructure and ecological fragmentation	Birds and avian diversity, mammals and small terrestrial fauna, pollinators and beneficial arthropods, forest-dependent species, dispersal-limited and habitat-dependent species, ecological connectivity and corridors, functional biodiversity and ecosystem structure
4. Habitat degradation and loss of environmental quality, including direct disturbance and establishment of alien species	Birds and avian diversity, insects and arthropods, flora and vegetation health, lichens, mosses, and bryophytes, soil biota and ecosystem quality

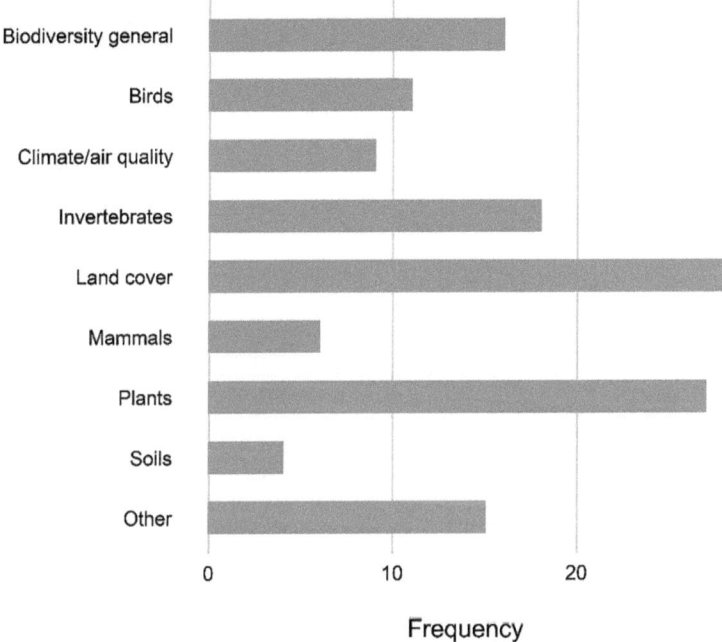

Fig. 2 The number of papers from the literature review addressing different response variable types in relation to threats to urbanization. N = 120 papers. Note that given paper can contribute to more than one response type (e.g., if both birds and invertebrates were considered in a single paper, both would be included in the respective categories). Authors' elaboration

considered, most (90.5%) studied terrestrial ecosystems, while 7% were focused on marine systems and only 2.5% on freshwater systems. The most frequent response variable type was land cover, followed by plants and general biodiversity. The most common animal group studied was invertebrates, followed by birds and mammals. There were only few studies specifically considering soils.

4 Discussion

4.1 Impacts of Urbanization on Species and Biodiversity Dynamics

Urbanization almost always causes a decrease in biodiversity compared to surrounding natural habitats [14] as a result of the multiple and interacting threats that impact on a large number of taxa [15]. The environmental changes caused by urbanization (Table 1) tend to create hostile conditions for many species, although there are some that can thrive in urban areas, often reaching densities that are much higher than in their natural habitats. These two groups of species are often respectively termed urban 'winners' and 'losers' [16]. Much research has been done on the characteristics that lead species to be winners or losers—in general, species with broad environmental tolerances, and (for animals) an omnivorous diet, tend to be winners, whereas specialist species tend to be losers.

Birds have been especially well-studied in urban environments, given that they are good indicators of environmental change [7], and there is much research on how individual habitat elements affect urban bird populations. The simplified habitat structure typical of urban environments in addition to increased disturbance levels excludes many species that have particular nesting or foraging requirements, species that typically include specialist species, such as birds of mature forest, or ground-nesting species [17]. For urban winners, however, urban habitats may provide favourable conditions that enable them to thrive in the city. For example, thanks to the urban heat island effect, whereby temperatures are higher in the city than the surrounding countryside [18], warmers winters may increase survival of resident species [19]; the few species that can nest in buildings may find safe refuges that mimic their natural habitats of cliffs and crags; and, the increased food available in urban areas, either through human waste or food intentionally provided (e.g., bird feeders), can provide a subsidy to omnivorous or granivorous species (e.g., Coccon and Fano [20]), increasing their body condition and survival prospects. However, this food subsidy may have negative effects on more specialist species [21] and may also benefit alien species disproportionately, which then have the potential to negatively affect native species through competition. Indeed, cities tend to be hotspots of alien species establishment [22, 23].

There have been some studies on individual species or species groups in an urban context in Italy [22–24], but there have been relatively few studies that have considered the role of key drivers of urbanization impacts. Fraissinet et al. [25] compared the bird community in Naples over time in a period of increased urbanization and loss of semi-natural habitat. Whilst they did not find evidence of changes in species richness, they found that the mix of species changed, with forest species tending to increase, but species of open and marginal habitats tending to decrease. Brambilla et al. [26] assessed the impact of restoration of peri-urban forest in Milan. They discovered that there were positive effects of ongoing and completed restoration on four species, and negative effects on only one (the Barn Swallow, a bird more associated with farmland). Interestingly, one of the positively affected birds was the Marsh Tit, a forest bird that does not commonly breed in urban areas. These two examples show the potential of habitat restoration in urban areas, but also that a focus on a single habitat type (e.g., forest) may not necessarily have wider benefits to biodiversity. Urban green spaces are known to be important habitats for urban biodiversity [27, 28], and their biodiversity is likely to be enhanced if they contain a diversity of habitats types.

Pollinators represent one of the most functionally important yet vulnerable groups of urban invertebrates, with global meta-analyses confirming widespread negative impacts of urbanization on both pollinator richness and abundance [29, 30]. Their dual habitat requirements for nesting sites and floral resources make them particularly vulnerable to multiple interacting urban environmental impacts [31, 32]. Similarly to birds, urban environments can create distinct groups of pollinator 'winners' and 'losers,' with Lepidoptera being the most severely affected group, and below-ground nesting solitary bee species also being particularly vulnerable [29, 30, 32]. Urban winners include generalist and large-bodied cavity nesting bee species and managed pollinators (e.g., honeybees), which can exploit novel urban resources [29, 32]. In general, non-native pollinator species richness increases with urbanization while native species richness declines, with cities serving as establishment hotspots for potentially invasive species [29, 33]. Urban losers comprise specialists with narrow habitat requirements, particularly ground-nesting solitary bees facing habitat loss through soil sealing and spring-flying species (e.g., Andrenidae) suffering from reduced early-season floral resources [29, 31, 32].

4.2 Landscape Drivers

The proportion of impervious surface emerges as an important landscape-scale driver, with studies consistently demonstrating negative relationships with pollinator abundance and richness [31, 32]. However, these effects can be mitigated when local floral resource availability is high, indicating that strategic enhancement of urban flowering plant communities can counteract landscape-scale pressures [29, 32]. Domestic gardens and allotments are potentially important urban habitats, supporting higher pollinator visitation rates despite comprising a tiny proportion of city area [32]. Research in northeastern Italy demonstrates that landscape heterogeneity and urban green space proportion positively influence pollinator abundance, though these effects primarily benefit generalists [34]. Ensuring that allotments and domestic gardens are protected in existing urban areas and included in future urban developments could benefit generalist pollinators and is likely to bring benefits for humans as well.

Italian urban planning initiatives increasingly recognize the need for additional green space in cities. Although not directly targeted at pollinators, the TALEA project in Bologna develops modular "Green Cells" to mitigate climate change [35] and inadvertently supports pollinator connectivity in the region. Furthermore, urban afforestation projects in several Italian cities incorporate "biodiversity strips" with perennial flowers and artificial nesting sites to support pollinators [36]. Management intensity proves critical for habitat quality, where reduced mowing frequency significantly increases pollinator abundance and richness. These simple interventions could provide substantial benefits given that lawns could comprise over 50% of urban green space in some cities [32].

4.3 Environmental Stressors and Plant Community Effects

Urban environments subject pollinators to multiple stressors beyond habitat loss. The urban heat island effect creates significantly warmer temperatures relative to surrounding landscapes, producing complex impacts on pollinator communities. While this can benefit some species through longer growing seasons and extended floral resource availability during periods when natural habitats offer limited resources, urban warming may also disrupt plant-pollinator interactions by shifting phenologies and facilitating establishment of non-native plants from warmer regions [31, 32].

Environmental contamination compounds these challenges. Neonicotinoid pesticides in ornamental gardens reduce colony growth and reproduction, while atmospheric pollution from exhaust fumes disrupts the chemical navigation systems pollinators rely on to locate flowers [32, 37, 38]. Urban beekeeping presents additional resource competition pressures, with evidence from Paris demonstrating that high honeybee colony densities negatively affect wild pollinator visitation within 500–1000 m of apiaries [32].

The dominance of exotic ornamental plants in urban areas fundamentally alters pollinator community structure. Although pollinators show no consistent preference for native over non-native plants when controlling for abundance [39], exotic species create temporal resource mismatches by retaining the flowering phenologies of their source regions rather than matching local pollinator emergence and activity patterns [31]. Despite these challenges, the strong positive relationship between flowering plant richness and pollinator diversity reinforces the critical importance of strategic urban plant community design [29, 31]. Surprisingly, while urbanization consistently reduces pollinator diversity, enhanced pollination services can occur in cities through abundant generalist and managed species [29], highlighting both the functional resilience of simplified systems and the conservation imperative to maintain diverse pollinator communities for long-term ecosystem stability.

4.4 Biodiversity Threats and Planning Responses

The threats to biodiversity identified through the literature review (as reported in the "Threat Category" column of Table 1) may be mitigated through targeted strategies and interventions to be integrated into the PGT, by translating the proposed guidelines into binding regulatory measures. The main strategies and interventions are outlined below, comprising both those drawn from the reviewed literature and those newly developed in the context of this study. Most of these actions aim to address more than one of the identified threats as detailed below.

The reduction of soil sealing, one of the main threats to biodiversity [40], has long been addressed in the implementation of Lombardy Regional Law no. 12 of 2005.

However, biodiversity objectives could also be more explicitly integrated into planning processes. Their integration could ensure a balance between development and ecological conservation, to be monitored over the long term, aligning spatial planning to the pillars of EU Green Infrastructure Strategy [41] and Nature Restoration Law [42]. In this context, the mitigation of the negative impact of artificial light at night is also recommended and should be included in sectoral plans, specifically by tuning the spectrum of LED lights to minimize exposure to blue light can reduce the attraction of nocturnal arthropods [43] or, more generally, by planning light intensity based on the ecological sensitivity of urban areas.

With specific reference to ensuring creation of urban habitats, support for pollinators, and enhancement of ecological connectivity for both animal and plant species, we propose to combine two actions identified in literature, i.e., planting native tree species [36] and retrofitting buildings with green roofs and walls in selected areas [44]. These areas should be identified based on (i) the current composition of urban green spaces; and (ii) an analysis of the impacts of urban heat islands on both the urban environment and human health, thereby prioritizing the shading of gathering places and the infrastructure connecting them. To operationalize the strategy proposed, municipalities could negotiate with private developers for the implementation of targeted public space improvements, such as the installation of green roofs, green walls, and tree planting [45], in lieu of standard development contributions.

Finally, with specific reference to the urbanized context of the Lombardy Region, which is predominantly composed of consolidated urban areas, strategies and interventions should be increasingly oriented towards the proper management of existing public green spaces, with the aim of supporting biodiversity by prioritizing quality over quantity. This measure has long been discussed in the literature [46], yet it is currently not widely implemented, neither in ordinary planning nor in sectoral plans. Indeed, at present, the Italian regional and national regulatory framework still tends to emphasize quantitative aspects over qualitative ones [46]. The regulatory framework was originally developed in the post war period, when cities were experiencing rapid growth, often in the absence of adequate regulation. During the second half of the twentieth century, lawmakers sought to manage this urban expansion by establishing quantitative standards for the provision of public facilities, including minimum per capita green space requirements [46]. Considering the quality of green spaces by incorporating vegetation standards into the plan (e.g., establishing a mix of trees, shrubs, and flowering meadows in designated areas planting native species) [36] could play a crucial role in enhancing urban biodiversity and mitigating the encroachment of alien species [47]. Prioritizing ecological quality and functionality could guarantee the alignment of Italian spatial planning to EU frameworks, such as the Nature Restoration Law [42]. This approach should also be applied to both public and private green spaces, the latter of which is typically overlooked in the literature, in order to ensure a consistent and integrated strategy for enhancing urban biodiversity.

5 Limitations and Future Pathways

Although restricting the literature review to the Italian context might appear to be a limitation, it can also be perceived as one of the study's strengths. This narrower scope allowed for realistic and targeted planning recommendations to be incorporated into municipal plans. The approach could be replicated at European or global scales by grouping studies by ecoregion, ensuring ecological consistency thoroughly investigating the structure of municipal plans across different countries. This will ensure that planning recommendations are both targeted and implementable, avoiding the generic guidelines that characterize the majority of the current scientific literature.

Acknowledgements This research was funded by the Italian National Recovery and Resilience Plan (NRRP), Mission 4 Component 2 Investment 1.4, Spoke 5—Call for tender No. 3138 of 16 December 2021, rectified by Decree n.3175 of 18 December 2021 of the Italian Ministry of University and Research funded by the European Union—NextGenerationEU. Award Number: Project code CN_00000033, Concession Decree No. 1034 of 17 June 2022 adopted by the Italian Ministry of University and Research, CUP, D43C22001250001 Project title "National Biodiversity Future Center—NBFC".

References

1. Seto KC, Guneralp B, Hutyra LR (2012) Global forecasts of urban expansion to 2030 and direct impacts on biodiversity and carbon pools. Proc Natl Acad Sci 109:16083–16088. https://doi.org/10.1073/pnas.1211658109
2. Beninde J, Veith M, Hochkirch A (2015) Biodiversity in cities needs space: a meta-analysis of factors determining intra-urban biodiversity variation. Ecol Lett 18:581–592
3. Simkin RD, Seto KC, McDonald RI, Jetz W (2022) Biodiversity impacts and conservation implications of urban land expansion projected to 2050. Proc Natl Acad Sci USA 119. https://doi.org/10.1073/pnas.2117297119
4. McKinney ML (2006) Urbanization as a major cause of biotic homogenization. Biol Conserv 127:247–260. https://doi.org/10.1016/j.biocon.2005.09.005
5. Faeth SH, Bang C, Saari S (2011) Urban biodiversity: patterns and mechanisms. Ann NY Acad Sci 1223:69–81. https://doi.org/10.1111/j.1749-6632.2010.05925.x
6. Reynolds C, Byrne MJ, Chamberlain DE, et al (2021) Urban animal diversity in the global south. In: Cities and Nature, pp 169–202
7. Sumasgutner P, Cunningham SJ, Hegemann A et al (2023) Interactive effects of rising temperatures and urbanisation on birds across different climate zones: a mechanistic perspective. Glob Chang Biol 29:2399–2420
8. Aronson MFJ, Lepczyk CA, Evans KL et al (2017) Biodiversity in the city: key challenges for urban green space management. Front Ecol Environ 15:189–196
9. Bush J, Doyon A (2019) Building urban resilience with nature-based solutions: how can urban planning contribute? Cities 95. https://doi.org/10.1016/j.cities.2019.102483
10. Baldock KCR, Goddard MA, Hicks DM et al (2019) A systems approach reveals urban pollinator hotspots and conservation opportunities. Nat Ecol Evol 3:363–373. https://doi.org/10.1038/s41559-018-0769-y
11. Punzo G, Castellano R, Bruno E (2022) Exploring land use determinants in Italian municipalities: comparison of spatial econometric models. Environ Ecol Stat 29:727–753

12. D'Onofrio R, Bocca A, Camaioni C (2025) Urban greening and local planning in Italy: a comparative study exploring the possibility of sustainable integration between urban plans. Sustain 17. https://doi.org/10.3390/su17073227
13. Zuniga-Palacios J, Zuria I, Castellanos I et al (2021) What do we know (and need to know) about the role of urban habitats as ecological traps? Systematic review and meta-analysis. Sci Total Environ 780:146559
14. Ibáñez-Álamo JD, Rubio E, Benedetti Y, Morelli F (2017) Global loss of avian evolutionary uniqueness in urban areas. Glob Chang Biol 23:2990–2998
15. Li G, Fang C, Li Y et al (2022) Global impacts of future urban expansion on terrestrial vertebrate diversity. Nat Commun 13:1628. https://doi.org/10.1038/s41467-022-29324-2
16. Alba R, Marcolin F, Assandri G et al (2025) Different traits shape winners and losers in urban bird assemblages across seasons. Sci Rep 15:1–11
17. Evans KL, Chamberlain DE, Hatchwell BJ et al (2011) What makes an urban bird? Glob Chang Biol 17:32–44
18. Oke TR (1982) The energetic basis of the urban heat island. Q J R Meteorol Soc 108:1–24. https://doi.org/10.1002/qj.49710845502
19. Chamberlain DE, Cannon AR, Toms MP, et al (2009) Avian productivity in urban landscapes: a review and meta-analysis. Ibis (Lond 1859) 151:1–18
20. Coccon F, Fano S (2020) Effects of a new waste collection policy on the population of yellow-legged gulls, Larus michahellis, in the historic centre of Venice (Italy). Eur J Wildl Res 66. https://doi.org/10.1007/s10344-020-01384-z
21. Shutt JD, Lees AC (2021) Killing with kindness: does widespread generalised provisioning of wildlife help or hinder biodiversity conservation efforts? Biol Conserv 261:109295
22. Mori E, Di Febbraro M, Foresta M et al (2013) Assessment of the current distribution of free-living parrots and parakeets (Aves: Psittaciformes) in Italy: a synthesis of published data and new records. Ital J Zool 80:158–167. https://doi.org/10.1080/11250003.2012.738713
23. Cardador L, Blackburn TM (2020) A global assessment of human influence on niche shifts and risk predictions of bird invasions. Glob Ecol Biogeogr 29:1956–1966
24. Orlando G, Chamberlain D (2023) Tawny Owl Strix aluco distribution in the urban landscape: the effect of habitat, noise and light pollution. Acta Ornithol 57:167–179
25. Fraissinet M, Ancillotto L, Migliozzi A et al (2023) Responses of avian assemblages to spatiotemporal landscape dynamics in urban ecosystems. Landsc Ecol 38:293–305. https://doi.org/10.1007/s10980-022-01550-5
26. Brambilla M, Foglini C, Vitulano S (2024) Small-scale forest restoration in peri-urban areas provides immediate benefits for birds. Bird Conserv Int 34. https://doi.org/10.1017/S0959270924000200
27. Ives CD, Lentini PE, Threlfall CG et al (2016) Cities are hotspots for threatened species. Glob Ecol Biogeogr 25:117–126
28. Lepczyk CA, Aronson MFJ, La Sorte FA (2023) Cities as sanctuaries. Front Ecol Environ 21:251–259
29. Liang H, He Y, Theodorou P, Yang C (2023) The effects of urbanization on pollinators and pollination: a meta-analysis. Ecol Lett 26:1629–1642
30. Fenoglio MS, Rossetti MR, Videla M (2020) Negative effects of urbanization on terrestrial arthropod communities: a meta-analysis. Glob Ecol Biogeogr 29:1412–1429
31. Harrison T, Winfree R (2015) Urban drivers of plant-pollinator interactions. Funct Ecol 29:879–888
32. Baldock KCR (2020) Opportunities and threats for pollinator conservation in global towns and cities. Curr Opin insect Sci 38:63–71
33. Fitch G, Wilson CJ, Glaum P et al (2019) Does urbanization favour exotic bee species? Implications for the conservation of native bees in cities. Biol Lett 15:20190574
34. Lorenzato L, Fantinato E, Sommaggio D et al (2024) Pollinator abundance, not the richness, benefits from urban green spaces in intensive agricultural land. Urban Ecosyst 27:1949–1959. https://doi.org/10.1007/s11252-024-01565-7

35. Roversi R, Longo D (2025) Regenerative and connective green cells to address fragmentation and climate change in cities: the TALEA project integrated solution. Sustain 17. https://doi.org/10.3390/su17073175
36. Resemini R, Geroldi C, Capotorti G, et al (2025) Building greener cities together: urban afforestation requires multiple skills to address social, ecological, and climate challenges. Plants 14. https://doi.org/10.3390/plants14030404
37. Girling RD, Lusebrink I, Farthing E et al (2013) Diesel exhaust rapidly degrades floral odours used by honeybees. Sci Rep 3:2779
38. Whitehorn PR, O'connor S, Wackers FL, Goulson D (2012) Neonicotinoid pesticide reduces bumble bee colony growth and queen production. Science 336(80):351–352
39. Williams NM, Cariveau D, Winfree R, Kremen C (2011) Bees in disturbed habitats use, but do not prefer, alien plants. Basic Appl Ecol 12:332–341
40. Arcidiaco L, Corongiu M (2025) Analysis of LULC change dynamics that have occurred in Tuscany (Italy) since 2007. Land 14. https://doi.org/10.3390/land14030443
41. European Commission (2013) COM(2013)249 Final. Communication from the Commission to the European Parliament, the Council, the European Economic and Social Committe and the Committee of the Regions. Green Infrastructure (GI)—Enhancing Europe's Natural Capital
42. European Union (2024) Regulation (EU) 2024/1991 of the European Parliament and of the Council of 24 June 2024 on nature restoration and amending Regulation (EU) 2022/869
43. Longcore T, Aldern HL, Eggers JF et al (2015) Tuning the white light spectrum of light emitting diode lamps to reduce attraction of nocturnal arthropods. Philos Trans R Soc B Biol Sci 370:20140125
44. Li H, Xiang Y, Yang W, et al (2024) Green roof development knowledge map: a review of visual analysis using CiteSpace and VOSviewer. Heliyon 10. https://doi.org/10.1016/j.heliyon.2024.e24958
45. De Toni A, Morello E (2025) Sustainable regeneration of industrial areas through enterprise engagement in collaborative design. J Clean Prod 144936
46. Rega C (2013) Ecological compensation in spatial planning in Italy. Impact Assess Proj Apprais 31:45–51. https://doi.org/10.1080/14615517.2012.760228
47. Norton BA, Bending GD, Clark R et al (2019) Urban meadows as an alternative to short mown grassland: effects of composition and height on biodiversity. Ecol Appl 29:e01946

Open Access This chapter is licensed under the terms of the Creative Commons Attribution-NonCommercial-NoDerivatives 4.0 International License (http://creativecommons.org/licenses/by-nc-nd/4.0/), which permits any noncommercial use, sharing, distribution and reproduction in any medium or format, as long as you give appropriate credit to the original author(s) and the source, provide a link to the Creative Commons license and indicate if you modified the licensed material. You do not have permission under this license to share adapted material derived from this chapter or parts of it.

The images or other third party material in this chapter are included in the chapter's Creative Commons license, unless indicated otherwise in a credit line to the material. If material is not included in the chapter's Creative Commons license and your intended use is not permitted by statutory regulation or exceeds the permitted use, you will need to obtain permission directly from the copyright holder.

Nexus Between Ecosystem Services Provision and Socio-Economic Variables: A Pathway for Equitable Planning

Silvia Ronchi and Marta Dell'Ovo

Keywords Environmental justice · Social inequality · Urban planning · Nature-based solutions · Planning parameters

1 Introduction

As major contributors to global greenhouse gas emissions, urban areas are among the most vulnerable to their consequences. The ongoing acceleration of urbanisation has intensified the exposure of cities to the adverse impacts of climate change, such as heatwaves, flooding, and prolonged droughts, which are increasingly intense and frequent. Moreover, projections estimate that by 2050, nearly 70% of the world's population will live in urban areas, with European countries expected to surpass this threshold significantly [1]. The expansion of urban areas inevitably leads to the reduction and fragmentation of natural capital (in the city and the surrounding areas), which, in turn, affects the quality and availability of ecosystem services (ES) [2] essential for maintaining urban liveability.

One approach to reduce the impact of increasing urbanisation is to minimise the spatial extent of urban areas and contrast "the sprawling layout of urban spaces, and the disordered use of land" [3], developing more compact city forms.

This approach can also significantly reduce green space availability within cities, with a consequent impact on ES provision associated with them [4].

S. Ronchi (✉) · M. Dell'Ovo
Department of Architecture and Urban Studies (DAStU), Politecnico di Milano, Milano, Italy
e-mail: silvia.ronchi@polimi.it

National Biodiversity Future Center (NBFC), Palermo, Italy

The crucial role of local green spaces is nowadays widely recognised in supporting both physical and mental well-being (e.g., [5, 6]). Furthermore, several ES offered by urban green spaces provide significant economic implications, both locally and regionally [7, 8], affecting house prices, the costs of buildings in terms of lighting, cooling, and heating systems, and attracting businesses and employees [9, 10]. Crucially, the spatial distribution and accessibility of urban green spaces raise essential questions of social equity, as unequal provision can exacerbate existing health and environmental disparities [11, 12].

Marginalised communities with less access to high-quality green spaces limit physical activity, mental well-being, and social interaction opportunities. Furthermore, areas with restricted availability of green spaces are more vulnerable to the effects of climate change, such as urban heat islands, while green infrastructure can help mitigate them.

In this sense, the concept of environmental justice encapsulates these inequities. Initially focused on the disproportionate exposure of marginalised communities to environmental hazards, the framework has evolved to include access to green spaces and ES as a fundamental dimension of justice [13, 14].

One empirical illustration of persistent inequity is the so-called luxury effect, whereby more affluent urban neighbourhoods benefit from significantly greater biodiversity and ES than less privileged areas [15, 16]. The luxury effect is driven by multiple socio-economic mechanisms, including greater private investment in landscaping, better maintenance of public green spaces, and historical planning biases that have prioritised affluent districts in infrastructure and greening policies [17].

Moreover, Leong et al. [16] underline how this phenomenon is a global dynamic, suggesting that the luxury effect is not a local anomaly but a consistent feature of urban ecosystems shaped by wealth and policy priorities. Addressing this issue requires urban planners to consider ecological equity as a central element of resilience and sustainability planning. Nevertheless, increased urbanisation trends have several potential societal benefits, which may lead to wider biodiversity benefits, as long as urban planning and management are driven by sustainability and social equity [18]. The luxury effect thus reinforces socio-spatial stratification, requiring a profound reorientation of urban planning toward ecological functionality, social equity, and participatory governance, ensuring that the environmental goods upon which all depend are equitably distributed and democratically governed so that cities remain habitable and fair amid demographic and climatic pressures.

This chapter aims to investigate the relationship between the spatial distribution of ES and socio-economic variables in dense urban areas to understand the dynamics of their degradation and the extent to which they are equitably accessible across different social groups within urban areas.

2 Methodology

Considering the definition provided by the Common International Classification of Ecosystem Services (CICES) [19], ES can be classified according to four main clusters consisting of Provisioning, Regulating and maintenance and Cultural and spanning from soil formation, fresh water, moderation of extreme events, aesthetic values, etc. The biophysical values, the economic, and the intangible ones provided by their presence can be estimated with the support of several methodologies and software that have been developed to support urban planning or policy decisions [20]. The economic benefits derive from the notion of Total Economic Value (TEV) [21], referring to the concept of use and non-use, and their estimation is based on two main categories of methods, revealed and stated preferences [22]. Moreover, several software programs were developed to integrate the quantification of biophysical and ecological benefits provided by ES and their evaluation and translation into monetary metrics. Among all of them, InVEST (Integrated Valuation of Ecosystem Services and Tradeoffs) [23] is an open-source software tool employed to spatially map and assess the value of ecosystem goods and services that support and enhance human well-being, evaluating how changes in land use or management, affect the delivery and value of ES. The outputs are georeferenced maps and quantitative estimates illustrating the supply, demand, and value of selected ES under different scenarios [24]. Socio-cultural dimensions, belonging to a perceptive and intangible sphere, are the most complex to be measured since non-material values such as emotional, affective, and symbolic aspects are not adequately captured by quantitative metrics, but they are better assessed through qualitative methods, participatory processes, and context-specific evaluative frameworks. Within this context, the Multicriteria Decision Analysis (MCDA) offers a robust framework for addressing the multidimensional nature of the problem [25], enabling the integration of diverse perspectives within the decision-making process [26].

Given these premises, the contribution proposes a multi-methodological approach (Fig. 1) to investigate the relationship between the spatial distribution of ES and socio-economic variables. Socio-economic data are elaborated considering the MCDA theory and entail selecting a consistent set of criteria based on the literature analysed on the luxury effect concept [15, 27–29], standardising the scores to be homogenous and then aggregating through the Weighted Linear Combination (WLC) and visualising a neutral scenario. The market value of residential buildings, the income of the population, and the percentage of elderly people (i.e., older than 65 years old) have been considered for the analysis. To evaluate the ES provision, the InVEST software allowed the elaboration of four services: the urban stormwater retention, the carbon storage, the habitat quality, and the urban cooling. In addition to reducing strongly correlated data layers that affect the final ES index's quality, a Principal Component Analysis (PCA) is suggested. PCA is a statistical technique used to reduce the dimensionality of a dataset while preserving as much variability as possible and supports interpretation by identifying the main axes of variation within the data [30, 31].

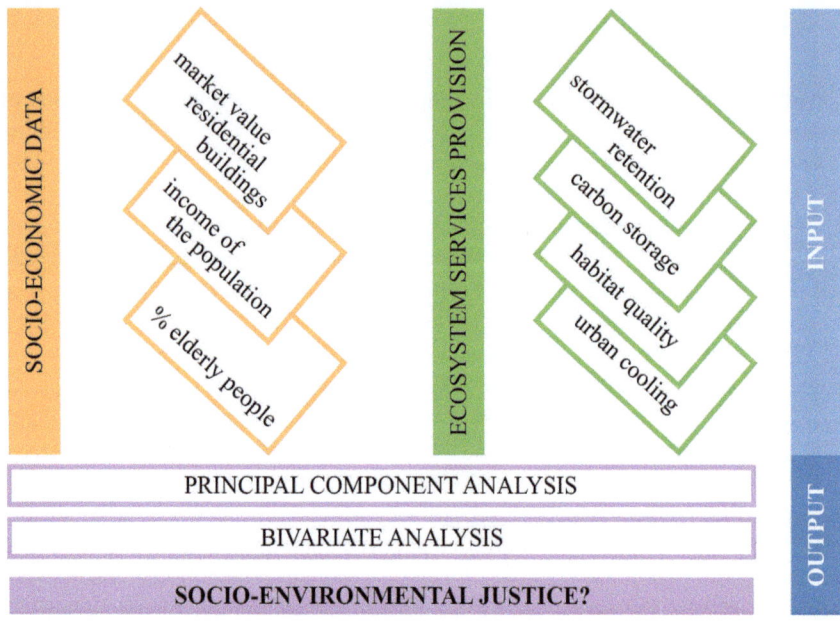

Fig. 1 Multi-methodological approach proposed

The two datasets, which are considered input data in our proposal, and consequently the two indices generated after the aggregation, have been developed independently so as not to overlap too many different layers, underestimate their importance, and lose the granularity of the data. To understand and analyse their potential correlation and relationship, the Bivariate Analysis is proposed [32]. Bivariate analysis is a statistical technique that measures the strength of association between pairs of variables (ES mapping and Socio-economic variables), assessing the overall power of a collection of variables to measure each other. Both datasets are georeferenced and visually represented by the support of maps and elaborated within the GIS domain (Geographic Information System). In fact, the bivariate analysis is a spatial technique used to examine the relationship between two variables across geographic space. It enables the visualisation of how two indicators co-vary within a defined area, helping to identify spatial patterns of association, correlation, or disparity between variables. The final output of the analysis will facilitate the examination of socio-economic and environmental dynamics within a specific geographic context, enabling the spatial identification and visualisation of patterns of socio-environmental justice or injustice, with the goal of informing urban planning and guiding equitable resource allocation.

3 Application in Two Italian Cities

The methodological approach was tested in two Italian cities, Turin and Milan, both located in the Po Valley in northern Italy. Despite sharing similar climatic challenges—such as air pollution and urban heat—they differ significantly in urban form and historical development.

Turin, capital of the Piedmont region, lies on the western edge of the Po Valley and has over 850,000 inhabitants (2025) across 130 km^2, making it Italy's fourth most populous city. Its metropolitan area presents a mix of urban and rural features, with a compact historical centre and increasingly fragmented patterns towards the outskirts. The eastern side transitions into low-density residential hills with green areas, while the western and southern sides are defined by dense neighbourhoods and industrial zones.

Milan, capital of the Lombardy region, is Italy's second largest city with more than 1.4 million inhabitants (2025) across 181 km^2. Its urban area extends beyond municipal borders, forming a continuous metropolitan region. Milan's northern and eastern sectors are predominantly high-density residential, while the southern and western areas include industrial, logistic, and newly developed zones. The city hosts a variety of green spaces, from large parks to local gardens.

4 Results

4.1 ES Assessment

Four ES mappings were conducted using InVEST and selected for their capacity to assess key aspects of urban ecological performance and to inform sustainable urban planning strategies, especially in critical, densely built environments like Milan and Turin. Habitat Quality (HQ) assesses the condition and spatial distribution of natural habitats by evaluating Land Use/Land Cover (LULC) concerning anthropogenic threats (i.e., industrial sites, infrastructural systems, agricultural areas), offering insights into biodiversity conservation and ecosystem health. HQ is considered a synthetic indicator for defining the ecological state of a region. This model was developed following the methodology outlined in Salata et al. [33] with the integration of NDVI data as suggested in Ronchi and Salata [34]

The Urban Stormwater Retention model estimates an annual water balance, i.e., the ability of different LULC types to retain and infiltrate rainfall, helping to understand areas more capable of reducing surface runoff and flood risk, especially relevant in highly sealed urban contexts.

Urban Cooling evaluates the ability of ecosystems to regulate temperature by identifying zones where vegetation mitigates the urban heat island effect, based on evapotranspiration and shading potential. This ES assessment is grounded in the framework developed by Ronchi et al. [35].

The last one, Carbon storage and sequestration model, quantifies the amount of carbon sequestered in different pools (aboveground and belowground biomass, soil, and dead matter), supporting climate mitigation analysis and planning of nature-based solutions in cities. Data and methods are derived from Arcidiacono et al. [36].

The ES mapping was represented using a hexagonal tessellation of 1 ha. Principal Component Analysis (PCA) was applied in both case studies as a multivariate statistical method to synthesise and interpret the spatial distribution of multiple ES. PCA aims to reduce data redundancy by transforming correlated ES layers into uncorrelated principal components, capturing the most meaningful variance. This improves composite ES maps' readability and contrast while giving ecological gradients a more accurate representation [37]. Results from the correlation matrix demonstrate a redundancy between HQ and UC (0.67) in the case study of Turin, while in Milan, a strong correlation emerges between HQ and UC (0.72) and between USWR and UC (0.68); therefore, it was decided for both cases not to consider the UC model (Fig. 2).

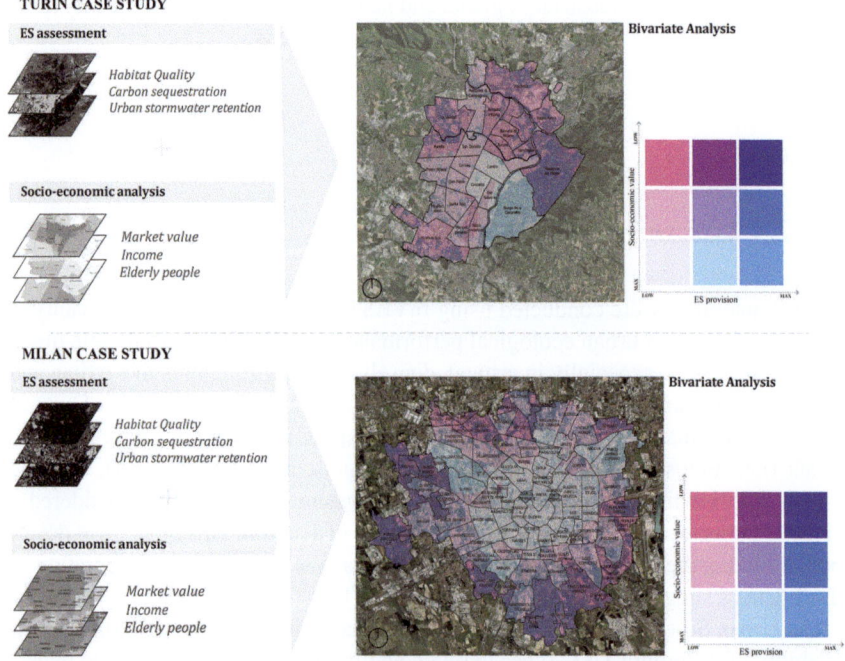

Fig. 2 Results from the bivariate analysis derived by ES assessment and Socio-economic data: above, the Turin case study; below, the Milan case study

4.2 Socio-economic Analysis

The criteria selected for the socio-economic evaluation were chosen after a comprehensive literature review focused on exploring the notion of the luxury effect and green gentrification within the urban context and, in detail, considering the correlation between socio-economic indicators and the concepts of ES and biodiversity [28, 29]. A frequency analysis has been developed to check the most pertinent criteria mentioned in the papers analysed and that are able to fit within the urban level. An additional filter has been set, which consists of being coherent in describing the Italian context and data availability. These phases brought the definition of three criteria:

1. The market value of residential buildings. Data have been collected considering the information provided by Agenzia delle Entrate and consisting of the average market value (€/sqm) of the civil dwellings in the normal state of preservation (abitazioni civili stato conservativo normale) for the second semester of 2024. The Agenzia delle Entrate defines homogeneous territorial areas nationwide to estimate average property values in Italy, and these are called OMI (Osservatorio del Mercato Immobiliare) zones. Each OMI zone represents an area of the city with similar urban and market characteristics. Within each zone, the Agency indicates minimum and maximum values for buying, selling and renting, broken down by property type (housing, stores, offices, etc.) and state of preservation. Data are provided two times per year.
2. The income of the population. Average income reflects the ability of households to access essential goods and services; a high income represents a higher level of quality of life. By comparing average incomes between different areas, spatial imbalances or social gaps can be revealed. For the city of Turin, data have been collected considering the official municipal geoportal and are divided by neighbourhoods. In Milan, data are divided by postal code, and the analysis is provided by the Ministry of Economy and Finance and elaborated by the national journal Corriere della Sera.
3. The percentage of elderly people (i.e., older than 65 years old). A high proportion of the elderly indicates an ageing population, which has direct effects on health care, welfare, the labour market, and social services. The proportion of elderly people also reflects demographic trends (low birth rate, increasing longevity) and may indicate areas at risk of depopulation or with low attractiveness to younger groups. Also in this case, for Turin, data have been collected considering the official municipal geoportal, while for Milan, considering the Nuclei di Identità Locale (NIL) information sheets defined by the Municipality of Milan.

Since the data for the different criteria were collected on varying scales, they were also represented, as in the ES mapping, using a 1-ha hexagonal tessellation through GIS operations.

4.3 A Bivariate Analysis

A bivariate analysis was conducted to establish the statistical association between the multifunctional ES and socio-economic factors represented on a cartographic basis and summarised in a 9-class chromatic matrix. The analysis highlights areas with low, medium, and high ES provision, which correspond to zones with different anthropogenic pressures (e.g., infrastructural systems and human activities) or a significant presence of natural and semi-natural elements that provide a vast amount of ES. The varying provision of ES is strongly connected to the urban design and development patterns of cities, directly influencing not only environmental dynamics but also residents' physical and psychological well-being through access to healthier and more resilient environments.

Similarly, socio-economic aspects are assessed along a gradient ranging from high to low, where "high" corresponds to high market value, a high percentage of elderly people, and high-income levels. In contrast, "low" refers to areas with low market value, a low percentage of elderly people, and low-income levels.

In the Turin case study, the historic centre's areas emerge as areas with reduced ecosystem capacity resulting from a deep "compact" urban basin, where an anthropic transformation is concentrated, creating a densely built urban core with a low permeability rate. At the same time, such areas combine a high socio-economic value, as they combine essential cultural and architectural value with excellent accessibility to services, making them attractive, exclusive, and prestigious residential areas. Their resident population often includes a higher share of elderly people and higher income brackets, thus having the economic capacity to afford to live in similar contexts.

Similar socio-economic conditions can be observed in the hillside areas of Turin (namely Borgo Po, Cavoretto, and Madonna del Pilone). In contrast, these areas differ significantly in terms of ES provision. The Turin hillside represents an important biodiversity reserve offering key regulating services, mainly due to its low-density and historically landscape-integrated settlement patterns.

A balanced condition is instead found in some peripheral neighbourhoods of the city, where both the ES and the socio-economic values are moderate. This is the case with Falchera and Regio Parco, both located on the northern outskirts of Turin. Falchera is characterised by low population density, with a prevalence of public housing and medium-rise condominiums. The quality of green spaces is discontinuous, but they have widespread and dense vegetation. Regio Parco, closer to the historic centre, has a higher population density, with mixed historic and popular buildings and green spaces that are qualitatively more structured but less extensive.

In the case study of Milan, the area characterised by low values of ES and high socio-economic indicators is significantly larger than that of Turin. This condition is not limited to the densely urbanised historic centre but includes a broader area that reaches the outer ring road. Two exceptions to this trend are the cases of Parco Sempione and Giardini di Porta Venezia, characterised by the predominance of urban green areas and therefore by high ES values.

A heterogeneous band of neighbourhoods extends around this central area, revealing contrasting relationships between socio-economic profiles and ES provision. On the one hand, neighbourhoods such as Dergano, Villapizzone, and Loreto are characterised by moderate socio-economic conditions, poor ES provision, high population density, and scarcity, both in quantity and quality, of green and permeable spaces. On the other hand, neighbourhoods such as QT8, San Siro, and Gallaratese show a markedly different trend: here, the presence of green infrastructures is dominant, and the overall permeability of open spaces is significantly higher.

The outer districts of the city, particularly those located in the southern area stretching from east to west and intersected by the Parco Agricolo Sud Milano, display a common condition: medium-to-high levels of ES alongside low socio-economic indicators. This pattern is primarily linked to low residential density and the predominantly agricultural character of these areas, which are especially suited to providing provisioning services.

This spatial heterogeneity emerged in the case studies highlights how urban morphology and land use patterns play a central role in shaping the distribution of ES in the metropolitan area, often reflecting past urban planning choices and settlement typologies.

5 Planning Perspectives

The analysis highlighted how the spatial distribution of ES and socio-economic variables within the urban areas of Milan and Turin is highly uneven, reflecting historical socio-territorial inequalities and reinforcing phenomena of spatial stratification. The observed relationship between ES provision and socio-economic conditions raises crucial questions for contemporary urban planning, calling for a reconsideration of the criteria and parameters commonly used to guide urban transformation processes.

Results demonstrate that it becomes essential to recognise "ecosystemic inequalities" as a structural dimension of urban environmental justice. It is crucial to adopt a planning approach that explicitly integrates ES considering what Dworczyk and Burkhard [38] define as "Service Providing Areas" (SPAs), "Service Demanding Areas" (SDAs), and "Service Benefiting Areas" (SBAs) to illustrate the different forms of demand that may emerge within a territory. Specifically, SPAs are defined as "areas where ES are provided," SDAs as "areas where people's needs or desires can be located," and SBAs as "areas where people knowingly or unknowingly benefit from the ES of interest".

This conceptual framework enables an urban planning model that is aware of and responsive to the interdependencies between natural capital and society. Therefore, it becomes more feasible to define context-sensitive minimum thresholds for areas capable of supporting ES provision, considering the specific socio-economic needs of each urban context.

The introduction of performance standards related to the presence and quality of ES, such as soil permeability, tree cover density, and green spaces, must be calibrated within an environmental justice perspective. This requires reformulating decision-making models to guarantee a fair distribution of environmental goods and greater community involvement in the prioritisation processes. Urban planning must therefore acknowledge that access to ES is a fundamental condition for urban liveability and collective health and well-being. Planning processes should also consider trends and future ES dynamics, assessing current patterns of ES provision and potential threats that could negatively impact such provision (e.g., land use changes) [39].

Finally, the proposed methodological approach confirms the value of integrated multidisciplinary tools in supporting decision-making processes to make political decision-makers informed, aware and responsible to orient urban transformations towards more equitable, resilient and sustainable models.

Acknowledgements This research has received funding from the Project "National Biodiversity Future Center—NBFC" funded under the National Recovery and Resilience Plan (NRRP), Mission 4 Component 2 Investment 1.4—Call for tender No. 3138 of 16 December 2021, rectified by Decree n.3175 of 18 December 2021 of Italian Ministry of University and Research funded by the European Union—NextGenerationEU; Project code CN_00000033, Concession Decree No. 1034 of 17 June 2022 adopted by the Italian Ministry of University and Research, CUP, D43C22001250001.

References

1. United Nations D of E and SAP division World Population Prospects 2024 (UN DESA/POP/2024/DC/NO. 10)
2. Bolund P, Hunhammar S (1999) Ecosystem services in urban areas. Ecol Econ 29:293–301. https://doi.org/10.1016/S0921-8009(99)00013-0
3. Jiang Y, Hou L, Shi T, Gui Q (2017) A review of urban planning research for climate change. Sustain 9. https://doi.org/10.3390/su9122224
4. Tratalos J, Fuller RA, Warren PH et al (2007) Urban form, biodiversity potential and ecosystem services. Landsc Urban Plan 83:308–317. https://doi.org/10.1016/j.landurbplan.2007.05.003
5. Chiesura A (2004) The role of urban parks for the sustainable city. Landsc Urban Plan 68:129–138. https://doi.org/10.1016/J.LANDURBPLAN.2003.08.003
6. Barton J, Pretty J (2010) What is the best dose of nature and green exercise for improving mental health—a multi-study analysis. Environ Sci Technol 44:3947–3955. https://doi.org/10.1021/ES903183R
7. Gregory McPherson E (1992) Accounting for benefits and costs of urban greenspace. Landsc Urban Plan 22:41–51. https://doi.org/10.1016/0169-2046(92)90006-L
8. Farber S, Costanza R, Childers DL et al (2006) Linking ecology and economics for ecosystem management. Bioscience 56:121–133. https://doi.org/10.1641/0006-3568(2006)056[0121:LEAEFE]2.0.CO;2
9. Morancho AB (2003) A hedonic valuation of urban green areas. Landsc Urban Plan 66:35–41. https://doi.org/10.1016/S0169-2046(03)00093-8
10. Kim HS, Lee GE, Lee JS, Choi Y (2019) Understanding the local impact of urban park plans and park typology on housing price: a case study of the Busan metropolitan region, Korea. Landsc Urban Plan 184:1–11. https://doi.org/10.1016/j.landurbplan.2018.12.007
11. Whitford V, Ennos AR, Handley JF (2001) "City form and natural process"—indicators for the ecological performance of urban areas and their application to Merseyside, UK. Landsc Urban Plan 57:91–103. https://doi.org/10.1016/S0169-2046(01)00192-X

12. Tzoulas K, Korpela K, Venn S et al (2007) Promoting ecosystem and human health in urban areas using Green Infrastructure: a literature review. Landsc Urban Plan 81:167–178. https://doi.org/10.1016/j.landurbplan.2007.02.001
13. Pearsall H, Pierce J (2010) Urban sustainability and environmental justice: evaluating the linkages in public planning/policy discourse. Local Environ 15:569–580. https://doi.org/10.1080/13549839.2010.487528
14. Jennings V, Johnson Gaither C, Gragg RS (2012) Promoting environmental justice through urban green space access: a synopsis. Environ Justice 5:1–7. https://doi.org/10.1089/ENV.2011.0007
15. Hope D, Gries C, Zhu W et al (2003) Socioeconomics drive urban plant diversity. Proc Natl Acad Sci USA 100:8788–8792. https://doi.org/10.1073/PNAS.1537557100/ASSET/9F5B8293-C983-4B01-B5B3-90DE21ECFEE5/ASSETS/GRAPHIC/PQ1537557002.JPEG
16. Leong M, Dunn RR, Trautwein MD (2018) Biodiversity and socioeconomics in the city: a review of the luxury effect. Biol Lett 14. https://doi.org/10.1098/RSBL.2018.0082;PAGE:STRING:ARTICLE/CHAPTER
17. Clarke LW, Jenerette GD, Davila A (2013) The luxury of vegetation and the legacy of tree biodiversity in Los Angeles, CA. Landsc Urban Plan 116:48–59. https://doi.org/10.1016/J.LANDURBPLAN.2013.04.006
18. Sanderson EW, Walston J, Robinson JG (2018) From bottleneck to breakthrough: urbanization and the future of biodiversity conservation. Bioscience 68:412–426. https://doi.org/10.1093/BIOSCI/BIY039
19. Haines-Young R, Potschin M (2010) Common international classification of ecosystem goods and services (CICES): consultation on version 4, August–December 2012. EEA framework contract no EEA/IEA/09/003. Contract 30. https://doi.org/10.1038/nature10650
20. Saarikoski H, Mustajoki J, Barton DN et al (2016) Multi-criteria decision analysis and cost-benefit analysis: comparing alternative frameworks for integrated valuation of ecosystem services. Ecosyst Serv 22:238–249. https://doi.org/10.1016/J.ECOSER.2016.10.014
21. Girard LF (1998) Conservation of cultural and natural heritage 25–50. https://doi.org/10.1007/978-94-017-1495-2_2
22. Hawkins K (2003) Economic valuation of ecosystem services. Univeryy of Minnesota
23. Sharp R, Tallis HT, Ricketts T, et al (2020) InVEST 3.7.0.post62+ug.h86a1183df108 User's guide
24. Burkhard B, Maes J (2017) Mapping ecosystem services. Pensoft Publishers, Sofia
25. Oppio A, Caprioli C, Dell'Ovo M, Bottero M (2024) Assessing ecosystem services through a multimethodological approach based on multicriteria analysis and cost-benefits analysis: a case study in Turin (Italy). J Clean Prod 472:143472. https://doi.org/10.1016/J.JCLEPRO.2024.143472
26. Greco S, Ehrgott M, Figueira JR (2016) Multiple criteria decision analysis. Springer, New York, NY
27. Aznarez C, Svenning JC, Pacheco JP, et al (2023) Luxury and legacy effects on urban biodiversity, vegetation cover and ecosystem services. npj Urban Sustain 3:1–11. https://doi.org/10.1038/s42949-023-00128-7
28. Dell'Ovo M, Datola G, Maiullari D, et al (2025) Green gentrification: a literature review of trends, challenges, and research opportunities. In: Gervasi O, Murgante B, Garau C, et al (eds) Computational science and its applications—ICCSA 2025 workshops. Springer Nature, Switzerland, pp 222–233. https://doi.org/10.1007/978-3-031-97645-2_15
29. Dell'Ovo M, Ronchi S, Regaiolo I, et al (2025) Planning for environmental justice. A multi-methodological approach. In: Gervasi O, Murgante B, Garau C, et al (eds) Computational science and its applications—ICCSA 2025 workshops. Springer Nature, Switzerland AG, pp 53–66. https://doi.org/10.1007/978-3-031-97589-9_5
30. Faisal K, Shaker A (2017) An investigation of GIS overlay and PCA techniques for urban environmental quality assessment: a case study in Toronto, Ontario, Canada. Sustainability 9:380. https://doi.org/10.3390/SU9030380

31. Salata S, Grillenzoni C (2021) A spatial evaluation of multifunctional ecosystem service networks using principal component analysis: a case of study in Turin, Italy. Ecol Indic 127:107758. https://doi.org/10.1016/j.ecolind.2021.107758
32. Guo D (2010) Local entropy map: a nonparametric approach to detecting spatially varying multivariate relationships. Int J Geogr Inf Sci 24:1367–1389. https://doi.org/10.1080/136588 11003619143
33. Salata S, Ronchi S, Arcidiacono A, Ghirardelli F (2017) Mapping habitat quality in the Lombardy region, Italy. One Ecosyst 2. https://doi.org/10.3897/ONEECO.2.E11402
34. Ronchi S, Salata S (2022) Insights for the enhancement of urban biodiversity using nature-based solutions: the role of urban spaces in green infrastructures design. In: Mahmoud IH, Morello E, de Oliveira F, Geneletti D (eds) Nature-based solutions for sustainable urban planning: greening cities, shaping cities. Springer International Publishing, Cham, pp 47–68
35. Ronchi S, Salata S, Arcidiacono A (2020) Which urban design parameters provide climate-proof cities? An application of the Urban Cooling InVEST Model in the city of Milan comparing historical planning morphologies. Sustain Cities Soc 63:102459. https://doi.org/10.1016/j.scs.2020.102459
36. Arcidiacono A, Ronchi S, Salata S (2015) Ecosystem services assessment using invest as a tool to support decision making process: critical issues and opportunities. Lect Notes Comput Sci (including Subser Lect Notes Artif Intell Lect Notes Bioinformatics) 9158:35–49. https://doi.org/10.1007/978-3-319-21410-8_3
37. Marsboom C, Vrebos D, Staes J, Meire P (2018) Using dimension reduction PCA to identify ecosystem service bundles. Ecol Indic 87:209–260. https://doi.org/10.1016/j.ecolind.2017.10.049
38. Dworczyk C, Burkhard B (2021) Conceptualising the demand for ecosystem services—an adapted spatial-structural approach. One Ecosyst 6:e65966. https://doi.org/10.3897/ONEECO.6.E65966
39. Loos J, Benra F, Berbés-Blázquez M, et al (2022) An environmental justice perspective on ecosystem services. Ambio 52(3):477–488. https://doi.org/10.1007/S13280-022-01812-1

Open Access This chapter is licensed under the terms of the Creative Commons Attribution-NonCommercial-NoDerivatives 4.0 International License (http://creativecommons.org/licenses/by-nc-nd/4.0/), which permits any noncommercial use, sharing, distribution and reproduction in any medium or format, as long as you give appropriate credit to the original author(s) and the source, provide a link to the Creative Commons license and indicate if you modified the licensed material. You do not have permission under this license to share adapted material derived from this chapter or parts of it.

The images or other third party material in this chapter are included in the chapter's Creative Commons license, unless indicated otherwise in a credit line to the material. If material is not included in the chapter's Creative Commons license and your intended use is not permitted by statutory regulation or exceeds the permitted use, you will need to obtain permission directly from the copyright holder.

//
Living in Harmony with Nature? Climate, Biodiversity and Planning Futures

Fabiano Lemes de Oliveira

Keywords Nature · Urban planning · Futures · Biodiversity · Climate change

1 Introduction

The two existential threats to humanity and the planet—the climate and the ecological crises—have generated narratives across a *hopefulness-despair spectrum*. On one end lies the catastrophe discourse, present since the early days of global warming awareness in the 1990s [1]. This narrative seeks to alert humankind to the disastrous consequences of anthropogenic action on the planet. Descriptions of scorched future landscapes, coupled with drastically reduced biodiversity, have populated dystopian imaginaries [2]. These visions portray futures marked by persistent crises and collapsed socio-ecological systems. While such visions warn of the impacts of current decisions, this approach has proven ineffective in causing behaviour change or in guiding transitions towards more desirable futures [3].

On the other end, positive narratives about building better futures for humanity and the planet have emerged as counterpoints. These visions of desirable futures function as instruments of hope intended to catalyse collective action towards common goals [4]. They counter the individual paralysis and disempowerment often induced by fear-driven climate discourses by mobilising collective hope around transformative and positive scenarios. This orientation is evident, for instance, in the UN's *Pact for the Future*, which explicitly rejects fear as a driver of change and instead affirms that "this is a moment of hope and opportunity" to build a world "in which humanity lives in harmony with nature" [5].

F. Lemes de Oliveira (✉)
Department of Architecture and Urban Studies, Politecnico di Milano, 20133 Milano, MI, Italy
e-mail: fabiano.lemes@polimi.it

© The Author(s) 2026
A. De Toni et al. (eds.), *Nature-Positive Cities: Adaptive Spatial Planning in Italy for an Ecological Urban Transition*,
PoliMI SpringerBriefs, https://doi.org/10.1007/978-3-032-06617-6_4

The future is continually in the making, and its openness invites consideration of possible, probable and preferable futures [6]. The latter are inherently normative: they reflect our values, expectations and desires of what *ought* to be. "Desirable" or "preferable" futures, in this sense, are those that increase the likelihood of effectively addressing the current crises and foster more balanced, synergistic and holistic relationships between humans and the rest of nature. Such visions provide directionality to and coordination of present-day actions, and can serve as powerful drivers of transformative change [7].

In order to build desirable futures, one must first be able to imagine them [8]. Such futures are often envisioned—explicitly or implicitly—in key global reports and policy frameworks. For instance, both the Kunming-Montreal Global Biodiversity Framework (GBF) [9] and the Pact for the Future envisage a future in which humans live "in harmony with nature", a vision that is also greatly manifested in the IPCC's Sixth Assessment Report (AR6) [10] through its Shared Socioeconomic Pathway SSP1 Sustainability.

This chapter examines two of these frameworks—the GBF and the IPCC AR6 report—to explore the foundations upon which they build positive futures for humans and nature in urban contexts. These documents have been selected for their global recognition and influence in shaping climate and biodiversity policy and action. The chapter first asks: what are the visions for urban biodiversity articulated in these frameworks? And, what forms of human-nature relationships do they prioritise? It then considers the implications of these visions for urbanism, reflecting how planning and design practices might evolve to operate within, and actively contribute to, nature-positive futures.

2 A Path for "Living in Harmony with Nature"? The Kunming-Montreal Global Biodiversity Framework

The Kunming-Montreal Global Biodiversity Framework, adopted in 2022 as the post-2020 global biodiversity framework, seeks to address the persisting loss of biodiversity driven by human activities. The GBF operates on two interlinked timeframes: a future-oriented medium-term vision for 2050, coupled with short-term actions to be taken by 2030. The overarching vision is for a world "in harmony with nature", in which biodiversity is "valued, conserved, restored and wisely used", [9] thereby maintaining and enhancing ecosystem services provision, and sustaining a healthy planet.

The GBF articulates four outcome-oriented goals for 2050. The first includes the maintenance, enhancement or restoration of ecosystem integrity, connectivity, and resilience, leading to a substantial increase in the extent of natural ecosystems. This is to be done in tandem with a halt in human-induced extinction of threatened species and the preservation of genetic diversity. The second goal centres on the sustainable use and management of biodiversity for the benefit of both present and future generations. The third focuses on the equitable share of genetic resources, while the last refers to the enabling conditions necessary to realise the previous goals, including financial resources, capacity building and international scientific cooperation.

These long-term goals are operationalised through a series of twenty-three action-oriented targets to be achieved by 2030. These include commitments to halt the loss of areas of high biodiversity importance and human-induced extinction of species, and restore at least 30 per cent of degraded terrestrial, freshwater and marine ecosystems. The GBF also calls for a significant reduction in pollution, a transition towards more biodiversity-rich agricultural practices, and the mainstreaming of nature-based solutions and other ecosystem-based approaches to support both people and nature.

Of particular relevance to urban contexts is the emphasis on enhancing the quality, quantity and connectivity of green and blue spaces. Furthermore, biodiversity and the multiple values of nature are to be integrated into planning policies, recognising cities as sites of biological recovery and stewardship. This includes a call for increase financial support to implement biodiversity strategies and plans at multiple scales.

The framework recognises that reversing biodiversity loss requires a multidimensional approach. It places value on the integration of diverse knowledge systems, including traditional and indigenous knowledge, as well as scientific evidence, to guide action. This epistemic inclusivity is accompanied by a normative commitment to intergenerational equity, emphasised by a call for youth engagement in decision-making, thereby aligning concerns for equity with democratic participation and long-term responsibility.

Yet, while the GBF calls for "transformative change", particularly through education aimed at shifting societal mindsets and behaviour, the precise meaning of "living in harmony with nature" remains somewhat under-defined. The GBF sits within the paradigm of "sustainable development". This view claims that economic development can be decoupled from environmental degradation and sustained over time, despite the confinements of the planet. The idea of sustainable and sustained development, albeit with ecological and environmental considerations, was codified in the Brundtland Report *Our Common Future* (1987) [11] and is anchored in the assumption of continuous economic growth as both feasible and desirable. This conceptual foundation stands in contrast to earlier critiques of growth-centred paradigms, such as those articulated by the Club of Rome's *Limits to Growth* (1972) [12], which challenged the viability of infinite growth on a finite planet [13]. Contemporary critiques—including the Planetary Boundaries framework [14] and most sharply the growing body of post-growth and de-growth scholarship [15]—further scrutinise the internal contradictions of the sustainable development model. They argue that the climate and biodiversity crises cannot be effectively addressed without rethinking the global economy's dependence on extractive and growth-driven logics [16]. Reflecting these concerns, in 2022 the Club of Rome launched a report to mark the 50th anniversary of *Limits to Growth*, in which it reiterates the need to transition from the current extractive economic system towards "wellbeing economies" that foregrounds the health of people and the planet.

In contrast, the GBF seemingly does not do away with the growth paradigm. Instead, it suggests that progress in "sustainable development" and in achieving the Sustainable Development Goals (SDGs) will create the necessary conditions for halting biodiversity loss and ultimately achieve "living in harmony with nature". The framework thus upholds a vision in which ecological integrity and development are not only reconcilable but mutually reinforcing.

The GBF projects a future in which humanity can better coexist with nature through a reformulated model aligned with the paradigm of sustainable development. Whether this is tenable remains a subject of debate, but the GBF undeniably articulates a global vision in which cities and other human systems are reimagined as active participants in sustaining the planet's biodiversity.

3 IPCC AR6 and the SSP1 Sustainability Pathway

The Intergovernmental Panel on Climate Change (IPCC) Assessment Reports are significant landmarks in the state-of-the-art of how science can point to the future consequences of actions and non-actions now. The Sixth Assessment Report (AR6) combined the Representative Concentration Pathways (RCPs), inherited from the Fifth Assessment Report (AR5), with Shared Socioeconomic Pathways (SSPs). While RCPs delineate possible greenhouse gas concentration trajectories and their associated radiative forcing levels by 2100—SSPs in turn offer internally consistent storylines of socioeconomic development, enabling a more integrated interplay between societal trends and climate outcomes. The links between climate change and biodiversity loss are presented prominently, with every increment in temperatures further exacerbating species extinction and ecosystems degradation.

The AR6 presented five SSPs across the spectrum of challenges to mitigation and adaptation: Sustainability (SSP1), Middle-of-the-road (SSP2), Regional rivalry (SSP3), Inequality (SSP4) and Fossil fuel-intensive (SSP5). These scenario storylines are long-term "what-if" descriptions of futures which today are deemed plausible [17].

Of the five SSPs outlined, SSP1 stands out as the most optimistic in addressing the twin crises of climate change and biodiversity loss. When paired with the lowest radiative forcing scenarios, SSP1-1.9 and SSP1-2.6, it projects futures where global warming is limited to 1.5 °C or 2 °C, respectively, by the end of the century. Crucially, this pathway is also the one most aligned with biodiversity protection and ecological regeneration.

The SSP1 world is one fundamentally reshaped by sustainability and sustainable development, which had undergone rapid transition to renewable energy sources and carbon–neutral economies [18]. SSP1-1.9 reaches net zero by around 2050, while SSP1-2.6 does so by 2075. Population growth is low, and significant reduction in inequalities and material consumption have been achieved. Cities are walkable, compact and resource-efficient settlements that minimize land-take, while simultaneously enhancing ecological functions. They are green—cities protect and enhance ecosystems in and around urban areas, further linking urban-peri urban and rural domains. Cities, in this scenario, do not only reduce their environmental footprint but actively contribute to ecological resilience through the integration of nature-based solutions (NBS).

It is a future in which a transformative change in the way humans approach and relate to nature has taken place, one involving synergistic relationships between anthropic and natural systems. Biodiversity conservation and ecosystem restoration are no longer marginal concerns but are positioned at the core of societal organisation, economic development, and climate policy. The co-benefits of addressing climate and ecological challenges together, with the understanding that biodiversity and healthy ecosystems are foundational to resilience, are embedded into policy, strategies and plans. Practices of afforestation and reforestation, for instance, are widely adopted, serving both as carbon sinks and as mechanisms to rebuild habitats and improve ecosystem connectivity.

SSP1 cities are nature-positive and mainstream a range of nature-based solutions across scales, including green corridors, green roofs, wetlands and parks [19]. These interventions are not only multifunctional in delivering ecosystem services such as urban cooling, air quality improvement, stormwater management, and food provisioning, but are also deliberately designed to enhance biodiversity. Besides, the report repeatedly stresses that nature-based strategies in cities must benefit nature, for instance in habitat creation and through ecological corridors. Furthermore, in being a low-warming scenario, SSP1 inherently reduces eventual impacts caused by increased temperatures on ecosystems.

Conclusively, SSP1 provides the most promising outlook for biodiversity. The pathway represents a shift in paradigm that would keep global warming within the limits of the Paris Agreement, benefiting humans and the more-than-human world. Although it does not frame it quite as clearly as the GBF, SSP1 is arguably a world in which a drastic shift in human-nature relationships has taken place—reconfiguring, as a consequence, socio-economic structures for the benefit of both human development and ecological integrity.

4 Planning Paradigms and Nature

Planning has long played a central role in forging human-nature relationships. Nature has consistently featured in planning imaginaries and interventions [20]. Historically, green space planning in industrialising cities responded to deteriorated environmental and health conditions—serving as counterpoints to the overcrowded, "miasmatic" urban environments of the time. They also became beacons for active recreation and sources of well-being. The idea of reconnection to nature has been, too, a key aspect of green space planning in cities. With the growth in urbanisation, access to nature and the countryside dwindled in many urban settings, especially in large agglomerations. In response, the planning of park systems—including for instance green belts, urban parks and green wedges [21]—became central attempts at re-naturing the urban fabric. The manners in which green space planning models were conceived and green spaces implemented in cities largely followed anthropocentric logics.

With growing concerns over global warming and biodiversity loss, the concept of green infrastructure gained prominence [22, 23]. It aimed to integrate ecosystem-services into planning discourse and practice, broadening the ecological rationale for green spaces. More recently, in the context of the aggravating crises, nature-based solutions [24] have emerged as a unifying framework encompassing a variety of ecosystem-based approaches to climate adaptation, ecosystem restoration and biodiversity support.

Planning has responded with embedding green infrastructure and NBS into its practices [25, 26], as well as broadening its underlying frameworks, as seen for instance in ecological urbanism [27], performance-based planning [28, 29] and biophilic approaches [30]—each seeking to forge more synergistic relationships between urban development and nature. As such, the aim of "living in harmony with nature" has not escaped planning considerations. Today, discussions broadening the recognition of the values of nature [31] have widened the epistemological base in planning [32], opening the spectrum of considerations beyond human-centred utilitarianism towards more inclusive understandings of multispecies urbanism [33] and more-than-human [34] planning.

Recently, the term "nature-positive" [35] has emerged, further signalling the need to shift from extractive to regenerative relationships with nature—going beyond protection of, and the principle of "do no harm", to one that actively enhances biodiversity and ecosystem vitality. Within this framing, nature-positive urbanism can be seen as a transformative approach to urban planning and design that actively enhances biodiversity, supports ecological integrity while enabling the flourishing of both human and non-human life.

Cities, despite their relatively limited geographic footprint globally, play a disproportionately large role in driving ecological overshoot. Yet, they also offer unique opportunities for transformational change. Their density, infrastructure, and flows of capital, life, and materials position them as potent laboratories for reimagining modes of coexistence.

To meet the challenges and opportunities outlined in the GBF and AR6 SSP1 scenario, planning must shift towards nature-based thinking—an approach that prioritises reciprocity, regeneration and long-term ecological health. It ought to be framed as a future-oriented, intergenerational and integrative endeavour, seeking to align urban development with ecological restoration and climate mitigation and adaptation. This shift is central to realising visions of living in harmony with nature—whether by 2050 or 2100—and demands a fundamental rethinking of the paradigms that have long governed urban transformation.

The juxtaposition of inherited paradigms, ambitious global frameworks, and emerging speculative signals reveals a contested space of transition in urban planning. To illustrate this, I adapt the *Futures Triangle* (Fig. 1) to map the current dynamics shaping the pursuit of "living in harmony with nature".

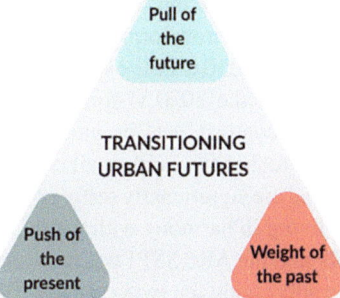

Fig. 1 A Futures Triangle for urban planning "in harmony with nature", mapping the tensions between legacy paradigms, present constraints, and emerging anticipations. Adapted from [36]

5 Discussion

The GBF and the AR6 both converge on the recognition of the interdependence of climate stability and biodiversity vitality. They ultimately recognize that a world in which humans flourish cannot be one in which biodiversity declines. In this shared understanding, both reinforce the need for urgent action: the GBF aims to halt and reverse biodiversity loss within this decade, while the AR6 warns that further delays in mitigation and adaptation efforts could foreclose the window of possibility of securing liveable futures.

While both frameworks envision more sustainable futures, they arguably diverge in the assumptions underpinning their respective transition pathways. The GBF frames its vision squarely within the paradigm of growth driven by the sustainable development agenda. It adheres to the assumption that growth can be decoupled from environmental and ecological degradation and biodiversity loss, despite increased claims that limitless economic growth is not possible on a planet with finite resources [37, 38]. The IPCC's AR6 SSP1 pathway, while also employing the language of sustainable development, opens conceptual space for transformative societal shifts beyond the growth paradigm foregrounding disruptive transitions. It envisions a future which could be potentially shaped by deep, systemic changes in economic structures—and compatible with wellbeing economy models [39] and post-growth thinking—even if these are not explicitly endorsed or articulated. As such, AR6 SSP1 signals toward a more radical reconfiguration of present-day and societal infrastructures.

These different conceptual underpinnings carry implications for the feasibility and meaning of the goal of living "in harmony with nature." In the GBF, this vision is set for 2050, with a strong emphasis on immediate action through 2030. While these short-term targets—focused on protection, restoration, and integration of biodiversity into policy—are laudable and aligned with the SDGs, they largely operate within an instrumental logic of managing nature and enhancing natural processes. Although the GBF acknowledges the need for a mindset shift, it falls short of proposing a pathway for systemic transformation in human–nature relations. In addition, the gap between 2030 and 2050 is little explored, with unclear milestones, which undermines the plausibility of the 2050 vision when assessed on the basis of the GBF alone. AR6 SSP1, by contrast, sets its most optimistic sustainability outcomes for 2100 (Fig. 2) and, if realised, would entail cities that are more ecologically integrated and landscapes that are significantly restored and diversified, suggesting a future in which the goal of living in harmony with nature might be more structurally embedded.

The GBF and the AR6 SSP1 pathway carry significant implications for the role of planning in addressing the twin crises of climate change and biodiversity loss. They prompt a critical reflection on how planning can serve as a transformative instrument for achieving the vision of living in harmony with nature. Such a transformation requires change in the deep structures of society. It also demands a reorientation of planning away from extractivism [40] and the pursuit of growth towards being an instrument of wellbeing for both humans and the more-than-human world. Planning needs to further develop integrative and systemic strategies and actions to support synergistic relationships between anthropic and natural systems. Besides, it ought to go further than what suggested in the GBF and the AR6 frameworks in terms of future-oriented thinking. Despite their ambition, these frameworks fall short of escaping the state-of-the-art of how to achieve the visions—largely recycling the promotion of nature-based solutions, green infrastructure and compact urban forms—without advancing exploration of novel spatial strategies, concepts and elements of future worlds that might be anticipated now, such as, for example, climate-resilient multi-species infrastructures and living materials. As such, high-level commitments and indications of desirable pathways have yet to be translated into visionary explorations of what a "living in harmony with nature" and SSP1-driven urban worlds might be like. Although these scenarios are important to the construction of sustainable and desirable futures, reframing current thinking is needed to advance imaginative, alternative and visionary solutions, ideas and strategies to contribute to resolving the current compounded challenges we face.

6 Conclusions

The chapter has examined how the vision of "living in harmony with nature" is articulated in two global frameworks—the Kunming-Montreal Global Biodiversity Framework (GBF) and the IPCC's Sixth Assessment Report (AR6)—and considered the implications for planning theory and practice. While both frameworks acknowledge the interdependence of limiting global warming and sustaining biodiversity, as

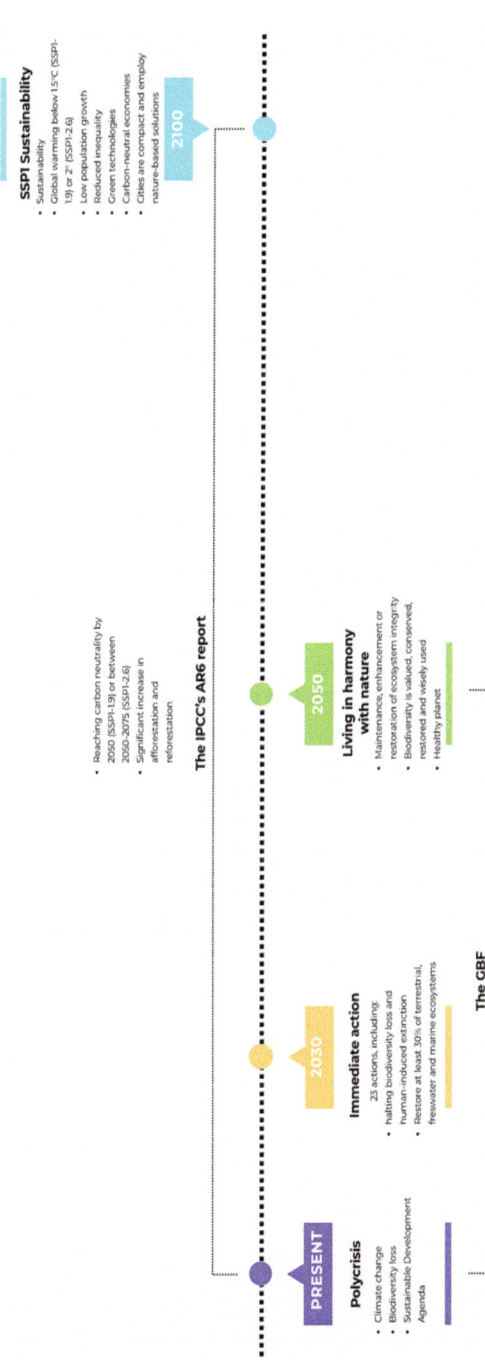

Fig. 2 Timeline of GBF and IPCC's AR6 SSP1 sustainability

well as the need for both a long-term vision and immediate action, they diverge in focus and in assumptions underpinning transformative change.

The GBF centres on halting and reversing biodiversity loss by 2030, and for humanity to live in harmony with nature by mid-century. In cities, it emphasises nature protection, restoration and connectivity. While it presents an actionable agenda aligned with the narrative of sustainable development, it falls short of offering a pathway for structural societal transformation. AR6 SSP1, in contrast, projects a long-term trajectory, envisioning 2100 as the horizon for a deeply decarbonised society and ecologically restored future. Its openness allows for the consideration of disruptive economic and societal transformations.

These carry implications to how achieving such futures are conceptualised and operationalised. While the GBF provides a concrete set of actions that would need to take place now, it lacks clear milestones between 2030 and 2050 casting doubt about how to move from existing development paradigms towards the visionary goal. In turn, the AR6 SSP1 suggests a more profound transformation of how lives are lived and of the cities that accommodate them in 2100. In both cases, the suggested nature-based interventions are capped within the limits of the state-of-the-art, without significant speculative exploration of approaches, strategies and types of solutions for biodiversity that could be anticipated from the future.

Planning must therefore navigate the tension between long-term aspirational goals and operational demands of immediate action. This requires an epistemic broadening of its scope. If planning is to contribute meaningfully to achieving harmony with nature it cannot be limited to embedding biodiversity and climate goals into existing systems. It must assume a more ambitious and anticipatory role—one that embeds nature-based thinking and reimagines human-nature relationships through the lenses of reciprocity, co-existence and multispecies flourishing. One of planning's responsibilities is to imagine these alternative futures, to explore visionary spatial imaginaries not yet fully captured in global frameworks. Another is to devise the mechanisms, strategies and enabling conditions for their realisation.

References

1. Soutar C, Wand APF (2022) Understanding the spectrum of anxiety responses to climate change: a systematic review of the qualitative literature. Int J Environ Res Public Health 19
2. Urry J (2016) What is the future? Polity Press
3. McPhearson T et al (2016) Positive visions for guiding urban transformations toward sustainable futures. Curr Opin Environ Sustain 22:33–40
4. Lemes de Oliveira F, Mahmoud I (2024) Desirable futures: human-nature relationships in urban planning and design. Futures 163:103444
5. United Nations (2024) Pact for the future, global digital compact and declaration on future generations. United Nations
6. Hancock T, Bezold C (1994) Possible futures, preferable futures. Healthc Forum J 37:23–29
7. Fedele G et al (2019) Transformative adaptation to climate change for sustainable social-ecological systems. Environ Sci Policy 101:116–125

8. Jasanoff S (2015) Future imperfect: science, technology, and the imaginations of modernity. In: Jasanoff S, Kim S-H (eds) Dreamscapes of modernity: sociotechnical imaginaries and the fabrication of power. University of Chicago Press, Chicago
9. Convention on Biological Diversity (2022) Kunming-Montreal Global biodiversity framework
10. IPCC (2023) Climate change 2023: synthesis report. Contribution of working groups I, II and III to the sixth assessment report of the intergovernmental panel on climate change. IPCC
11. Bruntland Report (1987) Our common future. World Commission on Environment and Development
12. Donella HM (1972) The limits to growth; a report for the club of Rome's project on the predicament of mankind. Universe Books, New York
13. Garforth L (2018) Green Utopias: environmental hope before and after nature. Polity Press
14. Kronenberg J et al (2024) Cities, planetary boundaries, and degrowth. Lancet Planet Health 8:e234–e241
15. Savini F (2025) Strategic planning for degrowth: what, who, how. Plan Theory 24:141–162
16. Otero I et al (2020) Biodiversity policy beyond economic growth. Conserv Lett 13:e12713
17. IPCC (2021) Climate change 2021: the physical science basis. Contribution of working group I to the sixth assessment report of the intergovernmental panel on climate change. Cambridge University Press
18. IPCC (2022) Climate change 2022: mitigation of climate change—working group III contribution to the sixth assessment report of the intergovernmental panel on climate change. IPCC
19. IPCC (2022) Climate change 2022: impacts, adaptation and vulnerability. Contribution of working group II to the sixth assessment report of the intergovernmental panel on climate change. Cambridge University Press
20. Pinder, D.: Visions of the City: Utopianism, Power and Politics in Twentieth-century Urbanism. Edinburgh University Press (2005)
21. Lemes de Oliveira F (2017) Green wedge urbanism: history, theory and contemporary practice. Bloomsbury, London
22. Benedict MA et al (2006) Green infrastructure linking landscapes and communities. Island Press, Washington, DC
23. European Commission (2010) Green infrastructure. European Commission
24. European Commission (2015) Towards an EU research and innovation policy agenda for nature-based solutions & re-naturing cities—final report of the horizon 2020 expert group on 'Nature based solutions and re-naturing cities'
25. Kabisch N, et al (2016) Nature-based solutions to climate change mitigation and adaptation in urban areas: perspectives on indicators, knowledge gaps, barriers, and opportunities for action. Ecol Soc 21
26. Mahmoud IH, et al (eds) (2022) Nature-based solutions for sustainable urban planning: greening cities, shaping cities. Springer, Cham
27. Mostafavi M, Doherty G (eds) (2010) Ecological urbanism. Lars Müller Publishers, Baden
28. Cortinovis C, Geneletti D (2020) A performance-based planning approach integrating supply and demand of urban ecosystem services. Landsc Urban Plann 201
29. Ronchi S, et al (2020) Integrating green infrastructure into spatial planning regulations to improve the performance of urban ecosystems. Insights from an Italian case study. Sustain Cities Soc 53:101907
30. Beatley T (2011) Biophilic cities: integrating nature into urban design and planning. Island Press, Washignton, D.C.
31. Lemes de Oliveira F (2025) Nature in nature-based solutions in urban planning. Landsc Urban Plann 256:105282
32. Pascual U et al (2023) Diverse values of nature for sustainability. Nature 620:813–823
33. Raymond CM, et al (2025) Applying multispecies justice in nature-based solutions and urban sustainability planning: tensions and prospects. npj Urban Sustain 5:2
34. Maller C (2021) Re-orienting nature-based solutions with more-than-human thinking. Cities 113

35. https://www.weforum.org/stories/2021/06/what-is-nature-positive-and-why-is-it-the-key-to-our-future/
36. Inayatullah S (2008) Six pillars: futures thinking for transforming. Foresight 10
37. Coscieme L et al (2020) Going beyond gross domestic product as an indicator to bring coherence to the sustainable development goals. J Clean Prod 248:119232
38. Hickel J, Kallis G (2020) Is green growth possible? New Polit Econ 25:469–486
39. Fioramonti L et al (2022) Wellbeing economy: an effective paradigm to mainstream post-growth policies? Ecol Econ 192:107261
40. Brenner N, Schmid C (2012) Planetary urbanization. In: Gandy M (ed) Urban constellations. Jovis, Berlin, pp 10–13

Open Access This chapter is licensed under the terms of the Creative Commons Attribution-NonCommercial-NoDerivatives 4.0 International License (http://creativecommons.org/licenses/by-nc-nd/4.0/), which permits any noncommercial use, sharing, distribution and reproduction in any medium or format, as long as you give appropriate credit to the original author(s) and the source, provide a link to the Creative Commons license and indicate if you modified the licensed material. You do not have permission under this license to share adapted material derived from this chapter or parts of it.

The images or other third party material in this chapter are included in the chapter's Creative Commons license, unless indicated otherwise in a credit line to the material. If material is not included in the chapter's Creative Commons license and your intended use is not permitted by statutory regulation or exceeds the permitted use, you will need to obtain permission directly from the copyright holder.

Ecological Connectivity Guides Spatial Planning

Alessandro Marucci◉ and Lorena Fiorini◉

Keywords Ecological connectivity · Land management policies · Spatial planning · Ecosystem functions and services · Biodiversity

1 Introduction

The growing knowledge of anthropogenic impacts on territory has highlighted the need to rethink the role of spatial planning. Traditionally understood as a tool for managing man-made transformations on the territory, planning must address the global challenges and play a central role in safeguarding biodiversity and promoting ecological connectivity.

Ecological connectivity is fundamental for maintaining biodiversity and its importance has been emphasised in several studies [1–3]. Currently, it constitutes a pillar of the EU Biodiversity Strategy 2030 [4], which aims to create a trans-European Network of ecological corridors.

The preservation of biodiversity inevitably involves the protection and enhancement of extensive areas characterised by high ecological value, such as parks and other protected areas. These areas are essential nodes within the territory, central to the survival of fauna and flora populations, in fact they act as sinks of biodiversity and starting points for ecological connectivity.

A. Marucci (✉) · L. Fiorini
Department of Civil, Construction-Architectural and Environmental Engineering, University of L'Aquila, 67100 L'Aquila, IT, Italy
e-mail: alessandro.marucci@univaq.it

L. Fiorini
e-mail: lorena.fiorini@univaq.it

© The Author(s) 2026
A. De Toni et al. (eds.), *Nature-Positive Cities: Adaptive Spatial Planning in Italy for an Ecological Urban Transition*,
PoliMI SpringerBriefs, https://doi.org/10.1007/978-3-032-06617-6_5

The 5th IUCN World Parks Congress, held in Durban (South Africa) in September 2003, was a milestone in the global conservation debate and introduced the concept of "Benefits Beyond Boundaries" [5]. The event fully recognised the invaluable role of parks and protected areas in safeguarding biodiversity and key ecosystems. However, it also raised significant discussions and reflections on the effectiveness of their use. Considering that many parks include large urbanised, industrial or intensive agricultural areas within them, deviating from the concept of high nature quality. The management of environmental conservation through core areas, i.e. areas with well-defined boundaries such as parks, is certainly easier. The boundaries are themselves a symbolic and practical barrier, harmonising the perception of the environmental qualities within them. This spatial delimitation also facilitates the investigation of naturalistic and physical aspects, as well as the definition of demographic and settlement dynamics, under the assumption that the phenomena are self-sufficient and closed with respect to systems outside them. Moreover, this approach makes governance and control actions easier, limiting the number of stakeholders and by centralising authority, partially releasing it from the complex local administrative dynamics [6].

Although this conservation approach through core areas, as an environmental enclave, has some limitations. In several cases, outstanding results have been achieved by removing large natural areas from irreversible episodes of alteration just through these conservation actions. However, the limitation of this approach is evident. In fact, there are many other areas of great ecological value, both nationally and internationally, that do not fit this group of protected areas and thus remain exposed to potential disruptive damage, with little possibility of safeguarding.

The strategic role of areas outside protected areas for the conservation of biodiversity, environmental continuity and the provision of ecosystem services has been a key topic in international debate for over thirty years. In fact, the Convention on Biological Diversity (CBD), signed in 1992 in Rio de Janeiro, introduced the Other Effective Area-based Conservation Measures (OECMs), which are specific geographical areas, different from the official protected areas, managed to ensure the long-term conservation of biodiversity. The IUCN World Commission on Protected Areas (IUCN-WCPA) is one of the six commissions of the IUCN, composed of experts in the management and conservation of protected areas. In 2019, the WCPA established a dedicated Task Force on OECMs, with the goal of defining and disseminating tools for the identification, management, and reporting of these areas. As mentioned, OECMs are distinguished by some basic characteristics. They are geographically defined areas with precise and recognised boundaries. In addition, they are subject to management/governance capable of producing, as a direct or indirect result, positive and durable effects for the conservation of biodiversity. This means that it is not necessarily the stated purpose of management that needs to be conservation but maybe, for example, territories dedicated to traditional productive activities, such as extensive pastoral farming or traditional fishing, in which long-established practices have allowed the maintenance of natural or semi-natural habitats, endemic species and functional ecological processes.

Since 1996, the Italian Institute for Environmental Protection and Research (ISPRA—Istituto Superiore per la Protezione e la Ricerca Ambientale), previously known as the Italian Environment Protection and Technical Services Agency (APAT—Agenzia per la Protezione dell'Ambiente e per i servizi tecnici), has actively promoted research on Ecological Networks in Italy. According to a growing awareness of the need for a dynamic approach to biodiversity conservation. This was a multi-year project carried out in collaboration with the National Institute of Urban Planning (INU—Istituto Nazionale di Urbanistica) aimed at defining tools and methodologies for the management of functional ecological connection areas, capable of guaranteeing ecological continuity between natural habitats based on the criteria of the Habitats Directive (92/43/EEC) and the Natura 2000 Project. Thus, awareness of the role of environmental continuity has inspired and still inspires debate on the definition of planning and management tools for governing ecological and ecosystem networks. But the technical implementation of planning designed for transformative spatial development in protection and conservation areas still reveals its partial effectiveness. A model based on static "zones and rules" has not enabled the enhancement of ecological connection areas due to the "in or out" theory.

With the aim of overcoming these limits, ISPRA itself is currently working on updating the Ecological Networks guidelines. This activity, developed in collaboration with the Department of Civil, Building, Architectural and Environmental Engineering (DICEAA) of the University of L'Aquila, is supported by a detailed analysis of the regulatory and/or planning instruments that the Italian regions have in place and carried out. This research aims to summarise the current status with respect to the implementation and management of ecological networks at a regional level, to understand what potential tools are currently available to deal with the issue of spatial fragmentation.

2 Three Core Concepts: Awareness, Consistency and Shared Responsibility

2.1 Awareness

Awareness of the role of biodiversity and ecological connectivity, including an understanding of their components such as ecosystem functions and services that contribute to human well-being, and the need to transfer this awareness from strategies into spatial and urban planning tool.

Awareness of the crucial role of biodiversity is particularly relevant in orienting planning and land management policies towards ecological connectivity. To do this, there must be a thorough understanding of the ecosystem functions and services provided by nature, recognising them as unavoidable elements in the quality of human life [7, 8]. This awareness must turn into an effective transfer of these concepts from international policies, which identify and define general strategies, to concrete spatial and urban planning instruments. European directives, such as the Habitats Directive (92/43/EEC) and the Birds Directive (2009/147/EC), already require the assessment of impacts on biodiversity, but more integration is needed at the planning level to design and implement effective ecological networks.

Despite the longstanding recognition of ecosystem connectivity as one of the key drivers of biodiversity, there is still ongoing search for a more inclusive and flexible approach compared to the insular model of protected areas. Major european strategies continue to emphasize the importance of promoting an interconnected ecosystem system by defining objectives and actions focused on maintaining ecosystem functions and services.

The EU Biodiversity Strategy for 2030, adopted in 2020 as part of the European Green Deal, underlines the need to restore ecological connectivity between natural and semi-natural habitats: "In order to have a truly coherent and resilient Trans-European Nature Network, it will be important to set up ecological corridors to prevent genetic isolation, allow for species migration, and maintain and enhance healthy ecosystems. In this context, investments in green and blue infrastructure should be promoted" [4]. A key objective is the establishment of a coherent and well-managed network of protected areas and Other Effective Area-based Conservation Measures (OECMs), complemented by green infrastructure and ecological corridors to enhance ecosystem resilience. Areas not under formal protection, i.e. OECMs, offer formal recognition to actors other than the State, including communities, local populations, and private entities that contribute significantly to environmental conservation.

The European Union Strategy on Adaptation to Climate Change [9] places strong emphasis on enhancing ecological connectivity to increase the capacity of ecosystems to absorb the impacts of climate change and to ensure the continuity of ecosystem services. Nature-based Solutions (NbS) are highlighted as one of the five key measures reported by Member States in the context of voluntary adaptation reporting: "Nature-based solutions encompass a range of ecosystem-based approaches that aim to increase resilience to climate change. … NbS are one of the five key type measures for adaptation in Member States' voluntary reporting on adaptation measures." [10] The strategy also stresses the importance of restoring urban green systems and improving connectivity between green spaces, encouraging european cities to develop ambitious urban greening plans that include parks, urban forests, green roofs and walls, hedgerows, meadows, and tree-lined avenues: "Nature-based Solutions for climate adaptation and disaster risk reduction can contribute to the EU nature restoration agenda. Applied at scale, they would enhance biodiversity in both urban and rural landscapes" [11].

The UN's 2030 Agenda, through its 17 Sustainable Development Goals (SDGs), recognizes the critical role of environmental connectivity in achieving global sustainability. It explicitly calls for the protection of the planet through sustainable consumption and production, responsible management of natural resources, and climate action. Several SDGs are directly linked to environmental connectivity and require coordinated efforts for effective implementation. Goal 11 (Sustainable Cities and

Communities) and Goal 15 (Life on Land) play a central role in promoting environmental connectivity, ecological networks, and the enhancement of ecosystem services. In particular, Goal 11 focuses on making cities and human settlements inclusive, safe, resilient, and sustainable. In this context, the presence of green and blue infrastructure, and their functional connectivity within urban and peri-urban areas [12–14], is essential to improve quality of life, public health, and resilience to climate change.

The New Urban Agenda [15] and the Global Biodiversity Framework [16] further reinforce this objective by highlighting the importance of linking green spaces to support urban biodiversity and ecosystem services. From an ecological perspective, the goal not only concerns the presence of green spaces but also emphasizes their quality, multifunctionality, and ecological connectivity, which are recognized as necessary conditions to ensure long-term benefits.

Awareness of the role of biodiversity and ecological connectivity is both well-established and persistent, including a nuanced understanding of their components as ecosystem functions and services that contribute to human well-being. The most recent global policy frameworks of the past decade have consistently emphasized the need to translate this awareness into spatial planning and urban development tools, continuing to promote Benefits Beyond Boundaries.

3 Consistency

After years of reflection and application practice, including a change in the scale of approaches, the concept of ecosystem continuity has undergone a deep evolution. Currently, this concept is an essential element of the Green Infrastructure model, which is defined, according to the European Commission, as: "A strategically planned network of natural and semi-natural areas with other environmental features designed and managed to deliver a wide range of ecosystem services. It incorporates green spaces (or blue if aquatic ecosystems are concerned) and other physical features in terrestrial (including coastal) and marine areas. On land, GI is present in rural and urban settings." [17].

Analysing the causes of the low efficiency of policies for the implementation of an ecosystem continuity system, there would be two fundamental issues: the first is related to the effectiveness of policies for the conservation and maintenance of biodiversity within protected areas, and the second is the absolute autonomy of soil transformation processes in the agricultural and semi-natural areas.

In the first case, an effort that has made it possible to preserve the environmental capital in the core areas, but which has not found an effective application in the areas designated to connect them. In fact, the implementation of a specific tool that can guarantee continuity passes, or should pass, through its translation as a strategic project of territorial planning outside these areas, for an effective ecological network. This involves its use as a benchmark in plan and project assessment processes (Strategic Environmental Assessment—SEA, Environmental

Impact Assessment—EIA and Impact Assessment—IA), through the enhancement of Protected Area systems and Natura 2000 Sites, and through the implementation of active interventions at the local scale [18].

Conversely, the second issue involves areas on which there are no strong protection and conservation regulations specifically aimed at maintaining ecosystem continuity. Europe has a long and consolidated experience of operative land planning at municipal level. However, regarding phenomena of environmental degradation and irreversible transformation of the landscape due to human actions, it is now showing the limits of an excessive decentralisation of decision-making autonomy. Soil transformations are always a response to simple local instances and beyond any strategic vision referred to higher governmental levels [19].

Moreover, although there have been numerous attempts to integrate ecological networks into the national and regional regulatory framework, the process is still open and heterogeneous. A survey of the regulatory instruments of the Italian regions on this topic, carried out by our research group as part of a project coordinated by ISPRA that is currently in progress, has shown that they have implemented, at various levels, models of ecological networks with extremely different timing and modalities. The legislative autonomy of the regions, despite some attempts with a national or subnational strategic approach (REN, EPA), has generated a heterogeneous picture that can be summarized in three main groups. The first is the environmental-landscape group where many regions have chosen to use Landscape Plans. The second group of a strategic-territorial type for which ecological networks are perceived as a structural element, with purposes of strategic control over land use and transformations. Finally, the third group of the instrumental-planning type where the ecological network is a characterizing and cogent element in urban planning.

Over time, it has become increasingly evident that the integration between environmental policies (from biodiversity conservation to the provision of ecosystem services) and land-use change policies is still neither efficient nor effective. Although the scientific community continues to propose up-to-date technical and administrative solutions, the integration between these spatial processes is still unsolved. The Italian regulatory framework on urban planning, anchored to Law 1150/42, is based on a static and non-resilient approach, unable to accommodate in a balanced and structural way the contents related to environmental sustainability, often influenced by the forecasting of economic development.

4 Shared Responsibility

International strategies, introduced and described before, define sustainability goals and fundamental criteria for ecological connectivity. These strategies are all characterised by a cross-border vision in which ecological coherence on a transboundary scale becomes valuable. Biodiversity loss in a single area can compromise the integrity of the entire ecological network, with consequences extending to even very distant territories. To avoid this, there is a need to improve policy coordination and effectiveness operate within both vertical and horizontal dimensions.

The vertical dimension concerns the complex administrative system consisting of different and overlapping levels such as regions, provinces, municipalities, but also unions of municipalities, metropolitan cities, mountain communities and parks. The high number of italian administrative levels does not facilitate coordination in the decision-making process that determines transformations and, consequently, increased pressures and threats on territories with high connectivity value. From the current trends, it does not seem possible to hypothesise a scenario that redefines the role of municipalities towards transformative decisions in a more cooperative and integrated direction, as indeed some european countries are doing. Rather, it is evident that the possibility of renewing decision-making mechanisms on the future of the territory, in a form compatible with local and strategic ecosystem arrangements, passes through an acquisition of awareness. Municipal administrations are lacking a marked power to direct coordination instruments, at least in areas not subject to environmental constraints.

This dimension of administrative structure shows little cogency especially in the parts of the territory with no environmental constraints. There is the certainty of a regulatory force of ecosystem constraints linked to specific protection rules, within protected areas. This contributes to a well-defined spatial determination of the configurations of land resources. A clear example of this is the phenomenon of land consumption in its different forms, including dispersed land consumption (urban sprawl, urban sprinkling), which is completely irrelevant in protected areas compared to the higher values of land consumption in the agricultural and semi-natural areas.

In addition to this, and regarding the horizontal dimension, there is the need for cooperation and shared vision across contiguous territories, regardless of their administrative boundaries. Italian municipalities enjoy almost complete autonomy in the management of urban and territorial planning, without being subjected to almost total strategic supervision, highlighting some of the most evident and negative consequences and proposing a path of at least partial recovery. This condition has gradually worsened over the course of several decades, in a climate of generalized indifference, but only recently have its pathological aspects been grasped: unjustified overurbanization and highly fragmented urban models, energy-intensive and destructive for ecosystems, in contrast with public interests in terms of environmental and urban quality [20]. The literature demonstrates the inadequacy of the current italian spatial planning system, centered on municipalities and substantially lacking in strategic vision. At least two aspects underline this pathology: first, the excessive diffusion of old municipal plans that for many years have not been able to gather the most recent territorial needs; second, the total and absurd independence of the urban evolution perspectives, always positive in the face of demographic dynamics, in some cases even drastically negative, and which nevertheless, in the medium-long term, determines serious phenomena of land consumption.

Overall, a sustainable approach requires that these authorities work in synergy and not in competition, recognising that spatial planning is itself a coordinative process that acts contextually on three axes: economic, environmental and social. In this framework, it is essential to identify territorial areas that are not just administrative but have a real physical and geographical coherence and promote shared strategic visions. The management of ecological connectivity at regional and interregional scales must be a shared responsibility among local governments, as land-use changes can have impacts that extend far beyond the areas where they occur.

5 Discussion and Conclusions

Thirty years after the Habitats Directive (92/43/EEC), and despite major changes in both the regulatory environment and global challenges, we can state that its basic goals remain undoubtedly relevant. They are based on universally recognised definitions and for this very reason have not changed over time. Today, these principles, which are fundamental for maintaining and improving ecological connectivity, are combined with new international strategies and innovative tools that make them an essential guide for spatial planning.

Therefore, rethinking governance models alone is not enough, as they must also consider the coherence of planning at different scales. Such land planning and management must, on the one hand, consider development opportunities and transformational needs, and on the other, ensure biodiversity conservation. This conservation, as already pointed out, in the introduction section, certainly implies the protection and enhancement of large areas with a high ecological value, but which must also integrate all the other areas that are fundamental to implementing an effective ecological network. At the same time, the need and capacity of territories to adapt to increasingly rapid and sometimes unpredictable phenomena must be considered. This requires an urgent move towards adaptive planning, which is not limited to identifying areas of intervention, but which is able to find and, above all, coordinate the various sources of funding (including extraordinary resources such as the National Recovery and Resilience Plan—NRRP) more effectively. The actions of different administrations on the territory do not always follow coherent views and consequently they may have potential negative impacts on the quality and/or ecosystem connection, in particular, when deriving from different funding sources. On the contrary, the real ecological connection that is naturally guaranteed also by residual areas should guide spatial planning choices towards a synergetic vision for an effective ecological network.

In the face of global challenges such as climate change, energy transition, socio-economic inequalities, loss of biodiversity and ecosystem fragmentation, it becomes imperative to establish broad and integrated governance frameworks, where different fields (urban, economic, environmental, social) interact continuously and spatial planning is able to respond in a timely and multi-scalar manner.

An important element for discussion, directly related to the three fundamental concepts described (awareness, consistency and shared responsibility), is the need to overcome the fragmentation of current competences and administrative tools. In fact, the urban planning and programming system cannot provide regulatory and operative containers into which new contents can be introduced. Too often, we see old plans that lack coordination and a unified strategic vision. Several municipal plans, elaborated over decades ago, are not structurally and operationally capable of implementing the most recent european and international policies on sustainability and ecological connectivity. In this context, public administrations and/or other stakeholders involved often use other more flexible devices, such as voluntary agreements [21, 22], because they are useful to overcome the current highlighted limitations.

For over thirty years, spatial planning has sought to define environmental continuity. It may now be time to reverse this perspective. Ecological connectivity guides spatial planning.

Acknowledgements This paper presents a reflection on ecological connectivity derived from our research activities developed within the work on updating the Ecological Network Italian guidelines, in collaborations with ISPRA and within the Integrated Project LIFE IMAGINE UMBRIA (LIFE19 IPE/IT/000015—Integrated MAnagement and Grant Investments for the N2000 NEtwork in Umbria).

References

1. Bennett AF (1999) Linkages in the landscape: the role of corridors and connectivity in wildlife conservation. IUCN, Gland, Switzerland and Cambridge, UK
2. Hilty JA, Lidicker WZ Jr, Merenlender AM (2006) Corridor ecology: the science and practice of linking landscapes for biodiversity conservation. Island Press, Washington, DC
3. Hilty J, Worboys G.L, Keeley A, Woodley S, Lausche B, Locke H, Carr M, Pulsford I, Pittock J, White JW, Theobald DM, Levine J, Reuling M, Watson JEM, Ament R, Tabor GM (2020) Guidelines for conserving connectivity through ecological networks and corridors. Best practice protected area guidelines series no. 30. IUCN, Gland, Switzerland
4. Commission E (2021) EU biodiversity strategy for 2030: bringing nature back into our lives. Publications Office of European Union, Luxembourg
5. IUCN (2005) Benefits beyond boundaries: proceedings of the Vth IUCN World Parks Congress, Durban, South Africa, 8–17 September 2003. IUCN, from https://portals.iucn.org/library/efiles/documents/2005-007.pdf. Accessed 12 June 2025
6. Romano B (2003) Il piano comunale strategico e i sistemi locali delle reti ecologiche: il tema dei corridoi. Ri-Vista. Res Landsc Architect 1:43–52
7. Daily GC, Alexander S, Ehrlich PR, Goulder L, Lubchenco J, Matson PA, Mooney HA, Postel, Schneider SH, Tilman D, Woodwell GM (1997) Ecosystem services: benefits supplied to human societies by natural ecosystems. Issues Ecol 2:1–18
8. Millennium Ecosystem Assessment (MA) (2005) Ecosystems and human well-being: synthesis. Island Press, Washington, DC
9. Commission E (2021) A new EU strategy on adaptation to climate change. European Commission, Brussels
10. European Environment Agency (EEA)—Climate-ADAPT. Nature-based solutions. https://climate-adapt.eea.europa.eu/en/eu-adaptation-policy/key-eu-actions/NbS. Accessed 12 June 2025
11. European Environment Agency (EEA) (2023) Scaling nature-based solutions for climate resilience and nature restoration (EEA Briefing No 08/2023). Publications Office of the European Union. https://www.eea.europa.eu/publications/scaling-nature-based-solutions. Accessed 12 June 2025

12. European Environment Agency (EEA) (2011) Green infrastructure and territorial cohesion. The concept of green infrastructure and its integration into policies using monitoring systems (EEA Technical report No 18/2011). Publications Office of the European Union. https://www.eea.europa.eu/en/analysis/publications/green-infrastructure-and-territorial-cohesion. Accessed 12 June 2025
13. Evans DL, Falagán N, Hardman CA, Kourmpetli S, Liu L, Mead BR, Davies JAC (2022) Ecosystem service delivery by urban agriculture and green infrastructure—a systematic review. Ecosyst Serv 54(2022):101405
14. Di Dato C, Pierantoni I, Fiorini L, Marucci A, Sargolini M (2024) Characterization of urban and peri-urban areas in Umbria region to identify their possible role in the conservation of natura 2000 network. In: Marucci A, Zullo F, Fiorini L, Saganeiti L (eds) Innovation in urban and regional planning. INPUT 2023. Lecture notes in civil engineering, vol 463, pp 659–668. Springer, Cham
15. United Nations (2017) New urban agenda. Resolution adopted by the General Assembly on 23 December 2016. A/RES/71/256. From https://docs.un.org/en/A/RES/71/256. Accessed 12 June 2025
16. Convention on Biological Diversity (2022) Kunming-Montreal global biodiversity framework. Decision 15/4. From https://www.cbd.int/doc/decisions/cop-15/cop-15-dec-04-en.pdf. Accessed 12 June 2025
17. European Commission (2013) Communication from the Commission to the European Parliament, the Council, the European Economic and Social Committee and the Committee of the Regions. Green Infrastructure (GI)—Enhancing Europe's Natural Capital (COM/2013/0249 final). Publications Office of the European Union, from https://eur-lex.europa.eu/resource.html?uri=cellar:d41348f2-01d5-4abe-b817-4c73e6f1b2df.0014.03/DOC_1&format=PDF. Accessed 12 June 2025
18. Lombardi L, Giunti M (2014) La traduzione della Rete Ecologica negli strumenti della pianificazione e nelle politiche di settore: dal sistema delle Aree protette al Piano paesaggistico regionale. In: Reti ecologiche e paesaggio per il governo del territorio in Toscana (a cura di Falqui e Paolinelli). Ed. ETS Pisa. Cap.9, pp 207–223
19. Vandelli L (2018) Il sistema delle autonomie locali. Il Mulino, Bologna
20. Romano B, Zullo F, Fiorini L, Marucci A (2019) Molecular no smart-planning in Italy: 8000 municipalities in action throughout the country. Sustainability 11(22):6467
21. Ciabò S, Romano B, Fiorini L, Marucci A, Olivieri S, Zullo F (2015) Parchi nella rete: l'accordo di varco. Reticula 9:8–15
22. Bastiani M (2013) I Contratti di Fiume come strumento di governance delle acque in ambito urbano. In (a cura di): ISPRA, Qualità dell'ambiente urbano—IX Rapporto—Focus su Acque e ambiente urbano. ISPRA 46/13, pp 59–63, Roma

Open Access This chapter is licensed under the terms of the Creative Commons Attribution-NonCommercial-NoDerivatives 4.0 International License (http://creativecommons.org/licenses/by-nc-nd/4.0/), which permits any noncommercial use, sharing, distribution and reproduction in any medium or format, as long as you give appropriate credit to the original author(s) and the source, provide a link to the Creative Commons license and indicate if you modified the licensed material. You do not have permission under this license to share adapted material derived from this chapter or parts of it.

The images or other third party material in this chapter are included in the chapter's Creative Commons license, unless indicated otherwise in a credit line to the material. If material is not included in the chapter's Creative Commons license and your intended use is not permitted by statutory regulation or exceeds the permitted use, you will need to obtain permission directly from the copyright holder.

Planning Ecological Networks from Regional to Local Level: Reflections to Support Biodiversity for People and Nature

Serena D'Ambrogi and Anna Chiesura

Keywords Ecological network · Territorial and spatial planning · Green infrastructure · Green urban planning

1 Introduction

1.1 Biodiversity Conservation and Spatial Planning: The Role of Ecological Networks

Ecological connectivity is defined as "*the degree to which the landscape facilitates or impedes movement among resource patches*" [18], or "*the ease with which individuals can move within the landscape*" [11]. It is a term that is regularly used for planning and conservation purposes [12] and most often refers to the potential movement of organisms.

The Intergovernmental Science-Policy Platform on Biodiversity and Ecosystem Services (IPBES) in the Global Assessment Report on Biodiversity and Ecosystem Services [9] identifies land-use change as the first direct driver of biodiversity loss, and stresses the necessity to provide planning options for effective conservation, restoration and sustainable use of nature and its ecosystem services. There is indeed a growing need to understand how to integrate biodiversity considerations into spatial and urban planning and to promote connectivity (structural and functional)

S. D'Ambrogi (✉) · A. Chiesura
ISPRA, Italian Institute for Environmental Protection and Research, Rome, Italy
e-mail: serena.dambrogi@isprambiente.it

A. Chiesura
e-mail: anna.chiesura@isprambiente.it

throughout the elements of the ecological networks (EN), with the aim to increase synergies between different types of land use, enabling to achieve multiple goals simultaneously.

In Italy, the debate on the relationship between EN and spatial planning has always been challenging. Even though biodiversity and conservation issues have increasingly influenced environmental policies and governance, regional planning experiences are limited to some landscape plans (Basilicata, Piemonte, Friuli Venezia Giulia, Toscana, Puglia, Abruzzo and Sardegna) and territorial plans (Veneto, Campania and Lombardia), as well as sectoral plans (Lazio). While this topic deserves special attention due to its potential to positively control the effects of land development, the relationship between EN and spatial and territorial planning appears to haven't been sufficiently explored. One reason for this may be the difficulty of innovating national and regional planning legislation. Although this has influenced national nature conservation policies through the introduction of ad hoc instruments, such as park plans (N.L. 394/91), the protection of biodiversity remains ancillary to the process of spatial planning and governance. Where dedicated normative provisions exist, they are often subjected to development interests or left to the discretion of local administrations at territorial and municipal levels. However, the need to strengthen the mainstreaming of nature and biodiversity issues into local planning instruments has been long debated and encouraged, with cities becoming increasingly relevant among institutional actors playing a key role [3, 19, 20].

1.2 The Multifunctional Ecological Network as Green Infrastructure Concept

The EN concept has been proposed as a useful mean to integrate biodiversity conservation into sustainable landscape development [15]. To be ecologically sustainable, landscape elements should support ecological processes to deliver ecosystem services to present and future generations. In multifunctional human-dominated landscapes, biodiversity conservation needs a coherent large-scale spatial structure of ecosystems, and EN can provide a critical framework for ecosystem design [5].

Especially in the context of both climate change and soil consumption challenges, the capacity of EN to meet biodiversity conservation objectives depends also on the degree to which it can support resilience within this appropriate matrix of landscapes, both in and between the core areas (as biodiversity main sources), thus enabling basic ecological processes and favouring biodiversity [1]. This network of features and functions could be planned and arranged for multiple goals, such as improving landscape quality and diversity, enhancing territorial resilience to climate change and support human health and wellbeing [4]. This acts as a multifunctional ecological network (MEN), aiming to connect ecosystems and regions [14] within a multipurpose ecosystem scenario to support sustainable territorial development [8].

MEN becomes a supporting infrastructure that provides more than one service/function within a wide area (region and/or province NUTS3 of EUROSTAT codes), combining global sustainability needs with local sensitivities and vocations [14]. It has to be seen as a structural and functional system able to support different regional and sectorial policies in a coherent framework to enhance synergies between them with the aim to improve environmental quality [14].

In recent years, it has become clear that natural and semi-natural spaces outside protected areas can, and do, contribute to reach biodiversity conservation while providing at the same time many other environmental and socio-economic services. Non-protected urban green areas, for example, play a crucial role in buffering the impacts of surrounding human development on natural protected areas,[1] essentially acting as a buffer zone by reducing direct pressures like habitat fragmentation and pollution. In addition, they help maintain landscape connectivity at various spatial levels, support species movement, and mitigate the effects of urbanization, while providing vital ecosystem services to humans (health and well-being, outdoor recreation, climate adaptation, cultural identity, etc.). Such multifunctionality and multiscalarity are the key qualities linking traditional EN with recent development in biodiversity and environmental policies, such as the concept of green and blue infrastructures (GBI). The European Commission defines GBI "*a strategically planned network of natural and semi-natural areas with other environmental features designed and managed to deliver a wide range of ecosystem services. It incorporates green spaces (or blue if aquatic ecosystems are concerned) and other physical features in terrestrial (including coastal) and marine areas. On land, this network is present in rural and urban settings*" [6].

The concept of GBI, pursuing multifunctional objectives, is expected to provide policy makers a wider range of policy options and implementation strategies to integrate ecological and sustainability concerns into spatial and territorial planning [7] in the rural and urban landscape. Thus, GBI represents an essential planning approach to enhance the quality of life while simultaneously protecting the natural capital, especially when it is currently not protected by national laws nor recognized by the planning systems [10] and strengthening resilience to climate change [17].

The multifunctional nature of GBI means that it delivers multiple services to meet multiple needs. The types of GBI required depend on the specific human and environmental contexts: inner cities, for example, may require green and blue spaces for recreation and regulation services, such as reducing the heat island effect and managing rainfall run-off; rural areas may require wilder habitats to improve connectivity between core wildlife areas such as Natura 2000 sites or buffering of agricultural land to reduce pesticide and fertilizers run-off or to improve pollination and sustainable pest control [10].

---

[1] More than 90% of the new soil consumed within protected areas is concentrated in national (almost 30 ha) and regional (34,11 ha) natural parks; the latter ones are those with the highest land consumption density (0,27 m^2 of new soil consumed per hectare of surface [16]).

2 Relationship Between the Different Planning Levels

A sound coordination between spatial planning instruments at different administrative levels (from regional to local/municipal) is crucial to guarantee the efficient implementation of EN as territorially open systems of relationships between biological and landscape components at several scales, each with a various mix of ecosystem services provided (from biodiversity conservation to climate adaptation). However, this highlights the need to develop strategic approaches to adaptive and multilevel planning policies that prioritise the protection of biodiversity, the enhancement of connectivity, and the fair and equitable provision of nature's ecosystem services.

The present contributions proposes a crossed analysis (Table 1) of two surveys carried out by ISPRA (Italian Institute for Environmental Protection and Research): one on the state of implementation of EN at the regional level (the administrative level of reference for spatial and landscape planning in the absence of national guidance), and the second on Urban greening plans (UGPs) of 10 Italian municipalities looking at the role of GBI and nature-based solutions within urban planning [2].

These two surveys provided an opportunity to verify how ENs are adopted within spatial and urban planning instruments in some regions, and whether and how the regional-level indications are translated into either binding or voluntary instruments at subordinate administrative levels. We have paid a particular attention to the role of the local administrative level in the belief that, as it will be argued later in the discussion, it can contribute to the planning of nature-positive cities and possibly influence ecological policies at higher levels through a bottom-up, vertical process.

2.1 The Regional Level

The ISPRA-University of L'Aquila survey, conducted between September 2024 and January 2025, has involved regional officials responsible for planning and environmental protection. The survey aimed to analyze the implementation of the EN concept in environmental and territorial governance instruments and regulations. While numerous good practices and positive examples were identified, the overall national picture highlights criticalities related to a lack of consistency in current regional EN designs. In particular, the planning and implementation of functional ecological connection elements is often conducted in the absence of multi-level, multi-sectoral and trans-regional coordination frameworks.

The need for a common approach has also been emphasized in the National Strategy for Biodiversity to 2030, which has given new impetus to the theme of connectivity and overcoming the critical issues mentioned above. This is achieved through sub-action A3, *"Ensuring the ecological-functional connection of protected areas at the local, national and supranational scale"*, which is part of objective A3.1.b) *"Updating and/or defining the regional ecological networks in line with the*

Table 1 A cross synthetic review of the two ISPRA's surveys: regional instruments (orange columns) and urban greening plans (green columns)

Region/Autonomous province (year of approval)	Regional norms for EN/indications for implementation at provincial and municipal levels	Municipalities (year of approval of UGPs)	UGPs—indications and proposal of natural/seminatural components of local/supralocal EN
Bolzano (–)	**There is no planned EN**	Bolzano (2022)	**No, but:** • Local landscape plan (Piano paesaggistico comunale) protects area of natural and landscape values, historical parks and villas, etc • Map of ENs and protected areas: periurban woods and Adige and Isarco rivers identified among strategic ecological corridors
Veneto (2020)	**Piano Territoriale Regionale di Coordinamento** …..the Provinces, the Metropolitan City of Venice and the Municipalities transpose the Regional Ecological Network into their territorial and urban planning instruments and adapt the regulations of the plans to this article, according to their respective competences, drawing inspiration from the principle of balance between the purposes of environmental valorisation and protection and economic growth [NT 2020, Art. 26]	Padova (2021)	**No, but:** • UGP provides in-depth analyses of local biodiversity (flora, fauna, protected areas. Urban parks and green public spaces) • Map with local ecological corridors and core areas

(continued)

Table 1 (continued)

Region/Autonomous province (year of approval)	Regional norms for EN/indications for implementation at provincial and municipal levels	Municipalities (year of approval of UGPs)	UGPs—indications and proposal of natural/seminatural components of local/supralocal EN
		Rovigo (2023)	**No,** but the Piano Urbanistico Comunale: • identifies a green belt project with the aim to conserve periurban high-natural values, potential core areas of territorial EN • proposes the institution of a locally protected area of relevant agricultural and landscape values (Parco agro-paesaggistico)
Piemonte (2017)	**Piano Paesaggistico Regionale** *Reference is made generically to subordinate (planning) instruments for the definition of rules of use for the elements of the Landscape Connection Networks*	Torino (2021)	**No,** but Turin's GI take part of a larger metropolitan EN with its network of urban green, periurban agricultural and hilly areas
		Vercelli (2021)	**No,** but areas of biodiversity conservation and restoration interests are identified such as the river Sesia and its riparian green areas, urban green network and agricultural areas

(continued)

Table 1 (continued)

Region/ Autonomous province (year of approval)	Regional norms for EN/indications for implementation at provincial and municipal levels	Municipalities (year of approval of UGPs)	UGPs—indications and proposal of natural/seminatural components of local/supralocal EN
Campania (2008)	**Piano Territoriale Regionale** *The Plan contains the following:* • *the drafting of Provincial Coordination Territorial Plans, paying specific attention to landscape* • *the construction of the RER (Regional Ecological Network), including sector planning and the construction of ecological networks at provincial and municipal levels*	Avellino (2023)	**Yes.** The local EN is composed of specific components acting as core areas, stepping stones of 1st and 2nd level, buffer zones and ecological corridors of territorial relevance. The municipal part of the river Fenestrelle is proposed as core area of EN (l.r.17/2003). Local EN with the urban GI through connections between higher-naturalistic agricultural and urban green areas
Toscana (2015)	**Piano Territoriale di Coordinamento a valenza paesaggistica** *Reference is made generically to subordinate (planning) instruments for the definition of rules of use for the elements of the Landscape Connection Networks*	Livorno (2023)	**No**, but: • the PSC (2023) has a map of local EN composed by woodlands, agro-ecosystems, river corridors and wetlands, coastal areas and urban EN (public green areas, relict natural areas, etc.); • Disciplina di piano_Invarianti strutturali indicates "urban river restoration as essential component of the urban green EN"

(continued)

Table 1 (continued)

Region/Autonomous province (year of approval)	Regional norms for EN/indications for implementation at provincial and municipal levels	Municipalities (year of approval of UGPs)	UGPs—indications and proposal of natural/seminatural components of local/supralocal EN
Emilia Romagna (–)	**There is no planned EN** *Regional Law no. 6/2005 defines that 'the provinces identify ecological connection areas, the conservation of which is regulated by general territorial (provinces) and urbanistic (municipalities) planning' (art. 7)*	Bologna (2021)	**No,** but: • UGP is part of the urban structural plan, which considers GBI as urban ecological network (ecorete urbana); • Rivers and hilly areas are considered as potentially functional for landscape and environmental connectivity within the provincial EN, as well as high-naturalistic public green areas

(continued)

Table 1 (continued)

Region/Autonomous province (year of approval)	Regional norms for EN/indications for implementation at provincial and municipal levels	Municipalities (year of approval of UGPs)	UGPs—indications and proposal of natural/seminatural components of local/supralocal EN
		Parma (2022)	**No,** but it refers to the general urbanistic plan: • The technical norms of the new Urban Structural Plan 2030 prescribe that every Piano urbanistico attuativo should include a map of EN, and that the new green spaces must relate to the EN (art.19); • The Urban structural Plan 2030 has produced a specific map of EN (1:25.000) composed by various green and blue natural and seminatural systems • Urban stepping stones, urban and periurban forestation areas, river areas, public and private green areas are mapped among the elements of the EN
		Forlì (2021)	**No,** but the UGP indicates that the new tree plantations and periurban forests will contribute to the supralocal EN

Note For editing reasons, information reported in the above table are extremely selected; refer to the complete document for a more complete picture *national strategic tool mentioned in point A1.3.a, and integrating them into territorial planning tools, particularly Regional Landscape Plans*".

The research process was complex, involving an in-depth analysis of existing EN in Italy. This analysis aimed to provide an updated and comprehensive overview of the territorial strategies adopted by regions for biodiversity management. A comparative analysis of the different regional situations revealed a wide range of approaches to defining and managing EN.

Of the six regions analysed for this cross-cutting analysis, it should be noted that, at the time of the survey, references to ENs in the Emilia-Romagna Region and the Autonomous Province of Bolzano were only present in normative documents, leaving autonomy to subordinate administrative levels (provincial and municipal) in terms of EN planning and governance. In the case of Emilia-Romagna Region, Regional Law no. 6/2005 defines that "*the provinces identify ecological connection areas, the conservation of which is regulated by general territorial (provinces) and urbanistic (municipalities) planning*" (Art. 7). The Metropolitan City of Bologna and the Province of Parma have then identified ENs in their spatial planning tools downstream of this law.

The other Regions (Toscana, Veneto, Campania and Piemonte) have instead incorporated ENs into territorial and/or landscape planning tools. Specifically, Toscana and Piamonte have included ENs in their Landscape Plans, and Veneto and Campania in their Regional Territorial Coordination Plans. In these cases, there are generic references to the implementation of ENs at subordinate administrative levels, covering all aspects of the Plans. These plans do not contain any specific tools that can univocally direct the implementation of what is foreseen and, in some cases, identified cartographically throughout the regional territory, thereby highlighting the strategic and guiding nature of this level of planning. Some more recent Landscape Plans (in the Friuli Venezia Giulia and Basilicata regions) took a more operative approach, as they contain specific guidelines for the implementation of regional ENs in subordinate plans. These guidelines ensure consistency and the effectiveness of regional indications.

2.2 Urban Greening Plans

While it is widely acknowledged that urbanization has severe negative effects on biodiversity, a rich and biodiverse nature and greenness can nevertheless exist in cities, with many ecological, environmental and socio-economic benefits for local populations. Cities have been increasingly recognized as crucial actors in biodiversity preservation and enhancement, as they can adopt policies to conserve and enhance local GBI and implement actions to improve landscape connectivity (see, for example, Target 12 of the Kumming-Montreal Global Biodiversity Framework). However, effective mainstreaming of biodiversity into urban planning is lagging behind, and no-mandatory national instruments exists to guarantee biodiversity conservation for nature-positive, healthy and resilient cities. UPGs, as called upon by both the European and national Biodiversity Strategy 2030, are deemed strategic—though voluntary—opportunities to systematically integrate nature and biodiversity into the medium-long term vision of structural urban plans.

With these premises, ISPRA has conducted an in-depth analysis of ten among the most recently (after 2020) approved Italian UGPs, with the aim to better understand their potential to advance local policies and urban planning towards more nature-positive cities. The 10 UPGs selected (Torino, Vercelli, Bolzano, Padova, Rovigo, Bologna, Parma, Forlì, Livorno and Avellino) have been red and interpreted through an ad hoc analytical framework, based on two representative reference guidelines defined at European (ICLEI Europe Guidelines) and national (Ministerial guidelines) levels. Five key macro areas have been identified as interpretative lens to analyse the heterogeneity of UPGs' contents and to answers specific research questions:

1. institutional context and normative framework: how the UGP interacts with the other local urban plans (the structural urban plan, in the first instance), which role does it play, and which relationships—if any—exist between urban and periurban GBI and EN at the various scales?
2. background knowledge of the territorial and environmental systems: which knowledge of the local green and natural systems is represented? Which components of the urban biodiversity are described, and which—if any—identified as strategical for municipal/provincial ecological network?
3. strategy, vision and objectives: which are the main goals of UGPs, and which strategies are defined to achieve them?
4. implementation actions and monitoring system: which actions are envisioned to implement the overall strategy and which monitoring system to verify their success?
5. public participation and awareness: how stakeholders and citizens have been involved in the UGP process?

To the present contribution, the first 3 macro areas are deemed of particular interest and will be further discussed.

The majority of UGPs analysed are approved as separated documents, with mainly strategic value for the structural/general urban plan, aiming at informing future developments with guidelines and indications to guarantee the right balance between GBI. Many UGPs have been written throughout consultation and dialogue with other relevant urban policy sectors (mobility, climate change, agriculture, etc.), therefore fostering the mainstreaming of urban greening and biodiversity into other urban planning procedures and practices. Most UGPs contain explicit references to provincial and/or regional territorial, biodiversity and landscape planning instruments, while less frequent is the coordination among the different administrative levels as far as the concept of EN is concerned. If almost all UGPs' visions are based on the efforts to connect urban and rural, GBI in the effort to minimize soils sealing, reduce urban fragmentation and increase ecosystem services for the health of people and nature, few of them propose a well-defined idea of local ecological network according to its traditional model (i.e., core areas, buffer zones, ecological corridors). Livorno, Parma and Avellino describe with specific cartography those natural and seminatural

Fig. 1 The urban trait of the Parma River (Photo by Anna Chiesura)

systems that—for ecological qualities, structures and functions—should be considered crucial components of the municipal EN. Nonetheless, almost all other UGPs contain in-depth details of local biodiversity and GBI, often indicating those of them having higher naturalistic or landscape values as potential components of EN at local or wider scales (see Table 1 for further details).

The background knowledge and the wealth of data and analysis (often at much finer scales and details) presented in the UGPs is of enormous value for well-informed policies and plans.

3 Discussion

The proposed crossed analysis of the two ISPRA's surveys allowed to investigate the relationship between EN/biodiversity planning instruments at the various administrative levels with the aim to verify if and how regional norms and indications are translated into either binding or voluntary (i.e., the UGPs) instruments at the local level. Results synthetized in Table 1 are here briefly reported and discussed.

The comparative analysis between the two planning normative frameworks (regional and municipal) regarding the implementation of EN, reveals that the regional level usually provides generic and not-binding indications for the lower administrative levels (Province and Municipalities). The regional level identifies *"areas with the highest biodiversity levels (i.e., natural parks, natural reserves, Natura 2000 sites) and the main drivers of ecological fragmentation (urban areas, roads and infrastructures, etc.) allowing to define the backbone of regional EN and highlighting sites that—because of their ecosystem intrinsic characteristics or their environmental functions—represent the reference frame for sustainable development scenarios"* (Regione Campania). Except in the case of Bolzano (an autonomous Province), in fact, regional norms and plans contain more (Veneto, Campania and Emilia-Romagna) or less (Piemonte and Toscana) explicit indications on how Provinces and Municipalities planning instruments should incorporate into their territorial and urban planning instruments the rules decided.

As far as the provincial level is concerned, for example, the role of EN is *"to contribute to building the EN connective tissue, through more targeted interventions, both in agriculture and urbanized areas"* (Regione Campania).

As far as the municipal level is concerned, the Regions of Veneto, Campania and Emilia-Romagna indicate that municipalities should identify in their urbanistic planning instruments measures to minimize impacts of anthropogenic pressures on ecological corridors, with the aim to guarantee their functional connectivity against naturals of artificial barriers.

Of the 6 Italian regions selected, two (Bolzano Autonomous Province, and Emilia-Romagna) do not have a planned EN at the competent administrative level. At the local level, however, UGPs identify valuable elements of local GBI and provide interesting references to potential natural and seminatural components for local/supralocal EN, such as:

- periurban woodlands, rivers' ecological corridors and historical parks and villas protected by the local landscape municipal plan (Bolzano);
- various green and blue systems identified by the urban structural plan, urban stepping stones, forestation and riparian areas, private and public green (Parma);
- new tree plantations and periurban forests (Forli');
- rivers and hilly areas, public green areas of high naturalistic values, local ecological network identified by urban structural plan (Bologna).

In these two cases, where no clear references to EN for biodiversity conservation exist at the competent regional level, UGPs represent a valuable contribution to fill this normative gap and to actively involve local planners and decision makers by committing them through a strategic—yet voluntary—plan to protect and enhance local biodiversity.

The analysis of the other 4 regions (Veneto, Piemonte, Campania and Toscana) highlights a picture where Territorial and Landscape Plans represent the normative references at regional level containing norms for the definition of regional EN and delegating subordinated (Piemonte and Toscana) and urban/urbanistic (Veneto and Campania) planning instruments for implementation and regulation. As for the local level, analysis of UGPs of the relative municipalities reveal an heterogenous picture, where:

- Avellino: a local EN is identified, composed of specific components acting as core areas, stepping stones of 1st and 2nd level, buffer zones and ecological corridors of territorial relevance. Local EN interacts with the urban GBI through connections between higher-naturalistic agricultural and urban green areas;

- no explicit local EN is identified but extensive, in-depth analysis are provided of local biodiversity and urban and periurban GBI; areas of high naturalistic and landscape values are proposed as potential components of the urban and supralocal EN (Padova, Rovigo, Torino, Vercelli and Livorno).

In these cases, similarly to the previous ones but in different normative contexts, the analysis of UGPs confirm the high potential of this voluntary sectoral planning instrument to inform local and supralocal nature and biodiversity conservation strategies. The often extensive and highly detailed analyses they provide of the local natural and seminatural systems, in fact, must be seen as a precious contribution to complete the "knowledge chain" that is necessary to implement EN, biodiversity and landscape policies at all administrative levels. The analytical and spatial details of UGPs maps provide crucial insights into the biodiversity profile of each single municipalities which is not available at higher level, and which therefore can be used to inform local and supralocal policies in a bottom-up decision-making process.

Furthermore, nature-based solutions envisioned in the 10 UGPs such as green roofs, rain gardens or urban allotments may represent additional stepping stones that act at the microscale (buildings, roads, neighbourhoods) to increase urban habitat connectivity [13].

Finally, it must be underlined that, for the purpose of this volume's nature-positive cities approach, a focus has been chosen on those aspects of UGPs concerning more directly biodiversity and nature conservation issues. Attention must be paid, however, on the fact that UGPs treats many other aspects of the urban fabric, intercepting almost all critical sectoral urban policies, from sustainable mobility and climate adaptation to urban regeneration and citizens' quality of life. That is to say that embedding nature-positive principles into structural planning means designing and implementing infrastructures that do not only enhances biodiversity, restore ecosystems and fosters human–nature connections, but also serves multiple social and economic goals.

4 Conclusions

We reflected on the roles of the different administrative levels in the effectiveness of actions for the implementation of EN, with a particular focus on the regional and local levels. The results of the analysis carried out and discussed in this paper show that, if regional planning tools define generic rules and concern strategic indications, the local level—through UGPs in our case—provides a multidisciplinary body of knowledge and a finer analysis of biodiversity in the increasingly urbanized pattern of contemporary societies. This can provide higher administrative levels with more detailed environmental and ecological knowledge.

We believe that the value of the UGPs can be realised more effectively through a circular, bottom-up governance model involving the different administrative levels. This circular model should define well-structured knowledge and monitoring systems that can assess the effectiveness of planned strategies and actions. In our opinion, this bottom-up process could facilitate a more adaptive and iterative approach to spatial planning, which is not currently possible due to the actual national planning normative framework. It could also complement the top-down implementation pathway by incorporating local visions, needs and information.

Analysis of the most recently approved UPGs has revealed that the concepts of urban GBI and ecosystem services are increasingly being adopted in local planning through strategic instruments that, although currently voluntary, can support the integration of biodiversity issues into policies that promote resilient and nature-positive cities. Making UGPs mandatory within the urban planning process could strengthen their crucial role in implementing local biodiversity through local EN.

References

1. Aubertin C, Rodary E (2011) Protected areas, sustainable land? Routledge, London
2. Chiesura A, et al (2024) I Piani comunali del verde: strumenti per riportare la natura nella nostra vita? Quaderni ISPRA Ambiente e Società 33/2024
3. CBD Secretariat of the Convention on Biological Diversity (2012) Cities and biodiversity outlook—executive summary. Montreal, 16 p
4. CEE (2009) WHITE PAPER adapting to climate change: towards a European framework for action COM (2009) 147/4. Brussels
5. D'Ambrogi S, Guccione M (2021) Multifunctional ecological networks as framework for landscape and spatial planning in Italy. In Catalano C, Leone M, Andreucci MB, Bretzel F, Menegoni P, Guarino R (eds) Urban services to ecosystems: green infrastructure benefits from the landscape to the urban scale. Springer Nature
6. EC European Commission (2013) Communication from the commission to the European parliament, the council, the European economic and social committee and the committee of the regions green infrastructure (GI) enhancing Europe's natural capital. COM/2013/0249 final
7. ESPON (2019) GRETA green infrastructure: enhancing biodiversity and ecosytem services for territorial development. www.espon.eu
8. Guccione M, Peano A (eds) (2003) Gestione delle aree di collegamento ecologico-funzionale Indirizzi e modalità operative per l'adeguamento degli strumenti di pianificazione del territorio in funzione della costruzione di reti ecologiche a scala locale APAT Manuali e linee guida 26/2003
9. IPBES (2019) Global assessment report on biodiversity and ecosystem services of the intergovernmental science- policy platform on biodiversity and ecosystem services. In: Brondizio ES, Settele J, Díaz S, Ngo HT (eds) IPBES
10. John H, Neubert M, Marrs C (eds) (2019) Green infrastructure handbook—conceptual & theoretical background, terms and definitions. First output of the Interreg Central Europe Project MaGICLandscapes—Managing Green Infrastructure in Central European Landscapes
11. Kindlmann P, Burel F (2008) Connectivity measures: a review. Landsc Ecol 23:879–890
12. Leitão AB, Miller J, Ahern J, McGarigal K (2006) Measuring landscapes: a planner's handbook. Island Press, Washington DC
13. Lynch A (2018) Creating effective urban greenways and stepping-stones: four critical gaps in habitat connectivity planning research. https://doi.org/10.1177/0885412218798334
14. Malcevschi S (2010) Reti ecologiche polivalenti. Il Verde Editoriale, Milano
15. Opdam P, Steingröver E, van Rooij S (2006) Ecological networks: a spatial concept for multi-actor planning of sustainable landscapes. Landsc Urban Plan 75(3–4):322–332

16. SNPA (2024) Consumo di suolo, dinamiche territoriali e servizi ecosistemici. Edizione 2024, Report ambientali SNPA, 43/2024
17. Suarez M, et al (2024) Urban resilience through green infrastructure: a framework for policy analysis applied to Madrid, Spain. Landsc Urban Plann 241
18. Taylor PD, Fahrig L, Henein K, Merriam G (1993) Connectivity is a vital element of landscape structure. Oikos 68(3):571–573. https://doi.org/10.2307/3544927
19. UN (2022) Cities and nature: planning for the future. White Paper. https://unhabitat.org/sites/default/files/2022/12/white_paper_cities_and_nature_rev2.pdf
20. UN (2023) UN habitat assembly adopted the resolution "Biodiverse and resilient cities: mainstreaming biodiversity and ecosystem services into urban-territorial planning", which explores how nature-based solutions can be used in combination with other types of interventions to generate multiple benefits for people and nature

Open Access This chapter is licensed under the terms of the Creative Commons Attribution-NonCommercial-NoDerivatives 4.0 International License (http://creativecommons.org/licenses/by-nc-nd/4.0/), which permits any noncommercial use, sharing, distribution and reproduction in any medium or format, as long as you give appropriate credit to the original author(s) and the source, provide a link to the Creative Commons license and indicate if you modified the licensed material. You do not have permission under this license to share adapted material derived from this chapter or parts of it.

The images or other third party material in this chapter are included in the chapter's Creative Commons license, unless indicated otherwise in a credit line to the material. If material is not included in the chapter's Creative Commons license and your intended use is not permitted by statutory regulation or exceeds the permitted use, you will need to obtain permission directly from the copyright holder.

Biodiversity and Landscape: Towards an Alliance in Italian Spatial Planning

Benedetta Giudice and Angioletta Voghera

Keywords Biodiversity preservation · Landscape planning · Ecological network · Green infrastructure · Local plan

1 Biodiversity and Landscape: Frameworks and Challenges in Europe and Italy

Numerous tools, methods, and strategies have emerged to foster sustainability and resilience in response to accelerating global changes, such as climate change, soil sealing, and pandemic-related crises. While many focus on technological innovation, they often remain fragmented. As a result, there is a growing call for integrated frameworks capable of addressing the complexity and dynamism of socio-ecological systems. Within this context, landscape and biodiversity have become central to advancing holistic approaches to socio-ecological resilience.

Approximately 80% of European habitats are currently degraded, underscoring the urgency for more ambitious and coordinated conservation policies. The targets set by the 1992 Convention on Biological Diversity (protecting 17% of terrestrial and inland water areas and 10% of marine areas by 2020) are now widely considered inadequate. In response, the Global Deal for Nature petition [1] and the

B. Giudice (✉) · A. Voghera
Interuniversity Department of Regional and Urban Studies and Planning (DIST), CED PPN, Politecnico di Torino, Torino, Italy
e-mail: benedetta.giudice@polito.it

A. Voghera
e-mail: angioletta.voghera@polito.it

Kunming-Montreal Global Biodiversity Framework call for more binding commitments, including legal restoration of at least 30% of all degraded ecosystems (terrestrial, water, marine, and coastal environments) and the conservation of 30% of land, waters, and seas.

Complementing these goals, the recently approved Nature Restoration Law, definitively adopted by the Council of the European Commission on June 17, 2024, reflects the urgency to address environmental degradation in Europe. As the first continent-wide legislative initiative explicitly dedicated to ecological restoration, the law marks a significant institutional commitment to advancing the European Green Deal goals and operationalizing the EU Biodiversity Strategy for 2030 by setting binding, time-bound targets: 60% of restoration actions must be completed by 2040, and 90% by 2050, with restored areas safeguarded against future degradation. By emphasizing integrating natural elements within urbanized contexts as essential for ecological transition and climate adaptation, the law mandates that member states adopt National Restoration Plans with clearly defined objectives and implementation measures.

Importantly, the law builds upon a long-standing urban and regional planning tradition. Pioneers such as Patrick Geddes, Ian McHarg, Frederick Steiner, Vittoria Calzolari, Roberto Gambino, Edoardo Salzano, and Alberto Magnaghi have promoted ecological approaches that "work with nature" through systemic landscape enhancement and context-sensitive urban design. Their legacy informs contemporary efforts to reintroduce permeable soils to rivers and floodplains, promote urban forests, deseal paved surfaces, and establish green and blue infrastructure. These interventions not only enhance biodiversity and ecological connectivity but also contribute to climate resilience by improving air, water, and soil quality and overcoming territorial vulnerabilities through a wide range of ecosystem services.

Since adopting the European Landscape Convention (2000), landscape has played an increasingly strategic role in broadening the scope of biodiversity policy. It enables a deeper integration between nature conservation and landscape planning. Indeed, as Gambino and Peano [2] argue, integrating these domains promotes a more holistic, place-based approach to biodiversity governance. In this context, landscape serves not only as a physical and ecological framework but also as a means of "discovery" [3] and spatial interpretation [4], providing a foundation for multi-scalar analysis and strategic design practices that align ecological, cultural, and spatial dimensions in planning processes.

In the Italian context, these two policy frameworks are effectively integrated, facilitating the development of strategies focused on conservation and territorial enhancement, with particular reference to regional landscape planning. Indeed, through the *Istituto Superiore per la Protezione e la Ricerca Ambientale* (ISPRA) and the Ministry of the Environment and Energy Security, Italy is advancing an integrated policy framework to meet the EU target of protecting 30% of territory by 2030. Central to this effort is reinforcing the national system of protected areas and integrating ecosystem conservation measures into spatial and landscape plans. These initiatives, which began over two decades ago with the development of the national ecological network guidelines [5], are now being updated. In parallel, green and blue infrastructure strategies [6] are being implemented through multiscale, multidimensional planning to enhance ecological and landscape biodiversity.

In this context, the recent amendment to Article 9 of the Italian Constitution[1] has brought renewed attention to the intricate interconnections among environment, territory, ecosystems, landscape, and cultural heritage. This constitutional innovation expands the roles and responsibilities of spatial planning across multiple governance scales. In particular, the "problematic triangulation between environment, territory, and landscape" [7] underscores the institutional complexities surrounding formulating strategic interventions. Within spatial planning, these dimensions often lead to differentiated trajectories and outcomes, shaped primarily by existing regulatory frameworks and the multiplicity of institutional actors involved. Alongside these themes, the notion of Nature, frequently framed through the conceptual lens of the environment [7], is increasingly emerging as a central construct within territorial, landscape, and environmental policies and strategies.

1.1 The Role of Regional Planning

The concept of ecological network, introduced in the 1990s to counteract habitat fragmentation and biodiversity loss, has become a cornerstone of Italian sustainability and land use policies [8]. Over time, it has gained growing significance in regional and provincial planning, supported by evolving regulatory and guiding frameworks. Nevertheless, while regional strategies play a key role, local implementation remains limited [9], hindering the translation of policy principles into effective urban development projects that conserve natural areas, reinforce ecological connectivity, and enhance the landscape. Moreover, persistent challenges also remain in grounding local actions within strong interpretative frameworks for biodiversity conservation.

Regional and landscape planning must increasingly broaden their scope to incorporate the emerging imperative of ecological transition and biodiversity conservation, enabling coordination across sectoral policies. While ecological connectivity remains a well-established focus, landscape planning has also broadened its scope of action and strategic functions to engage with new priorities such as climate change adaptation, risk and vulnerability management, and territorial resilience. Despite this, challenges persist in delivering design quality and integrated implementation. Similarly, protected area planning has evolved to support biodiversity as a foundation for sustainable development and territorial transformation. However, this ambition continues to be constrained by the rigidity of an outdated legislative framework (Law No. 394/1991), which no longer reflects the complexity of current challenges.

---

[1] The Republic promotes the development of culture and scientific and technical research. It safeguards the landscape and the historical and artistic heritage of the Nation. It protects the environment, biodiversity, and ecosystems, also in the interest of future generations.

Moreover, through various interactions between land-use and landscape policies, regional planning serves as a strategic intersection of environment, territory, landscape, and nature. In Italy, the adoption of the European Landscape Convention in 2000 (ratified in 2006) together with the approval of the *Codice dei Beni Culturali e del Paesaggio* (henceforth "the Code") in 2004, revitalized landscape planning, spurring innovation across regional, intermediate, and local planning levels. A key outcome has been the integration of strategic spatial development components with landscape protection. The regional landscape plan (*Piano Paesaggistico Regionale*—PPR), in particular, has emerged as a systemic tool that links regulations, policies, and strategic vision, fostering forward-looking ecological enhancement and experimental responses to climate change adaptation and landscape vulnerabilities mitigation [10]. Nevertheless, regional landscape planning in Italy remains fragmented yet continuously evolving [11]. To date, seven PPRs have been approved in line with the Code (Sardinia—limited to the coastal zone, Tuscany, Puglia, Piedmont, Friuli Venezia-Giulia, Lazio, and Basilicata, while Veneto and Lombardy have adopted their PPRs. Ten Regions are currently drafting or revising their PPRs, including Abruzzo, Calabria, Campania, Emilia-Romagna, Liguria, Marche, Molise, Umbria, and Sardinia (for inland areas). Finally, adapting local planning tools to PPRs also proceeds unevenly, reflecting regional disparities in implementation pace.

2 Multi-scalar Governance of Biodiversity and Landscape in the Piedmont Region

2.1 The Regional Framework

The Piedmont Region is undergoing major planning transformations. Since its current Regional Territorial Plan (*Piano Territoriale Regionale*—PTR) was approved in 2011, the region has progressively revised sectoral and non-sectoral planning tools. Piedmont has also recently approved two key strategic frameworks: the Regional Strategy for Sustainable Development and the Regional Strategy on Climate Change, following a complex process to deliver integrated technical, operational, and regulatory responses to contemporary challenges.

Although formally distinct, the PTR and the PPR have been built since the mid-2000s from a common cognitive framework, strategic vision, general objectives, and the environmental assessment process. This alignment ensures strong regulatory coherence, based on five core policies: (i) territorial regeneration and landscape protection and enhancement; (ii) environmental sustainability and energy efficiency; (iii) territorial integration of mobility, communication, and logistics infrastructures; (iv) research, innovation, and economic-productive transition; and (v) the enhancement of human capital and institutional capacities. While they share a strategic foundation, their specific objectives and actions differ, reflecting their distinct scopes.

The PPR of Piedmont, initiated in 2009 and approved in 2017 through an agreement with the former Ministry of Culture, establishes key sustainability principles: responsible land use, reduced agro-natural soil consumption, and protection of landscape features that constitute the identity of each territory. These aims are pursued through knowledge-based actions for protection, enhancement, and regulatory guidance. The PPR's strategic apparatus supports spatial planning and major projects such as UNESCO and MAB sites, the Corona Verde initiative, river and lake contracts, the regional ecological network, and land take monitoring. Its regulatory apparatus, meanwhile, guides the management of landscape assets, informs project authorization, and assists local landscape commissions.

The interpretive and strategic framework of Piedmont's diverse landscape is articulated through the identification of 76 landscape areas (*ambiti di paesaggio*). These territorial units are recognized and described through detailed fact sheets highlighting their structuring, qualifying, and characterizing features. The landscape areas serve not only as a reinterpretation and synthetic representation of the territory but also as the foundational framework for defining landscape quality objectives aimed at protecting and enhancing the territory.

The environment, shaped over millennia through human-nature interaction, is a structuring element of the regional landscape. This relationship underpins the concept of the landscape connection network, a strategic scenario of the plan integrating protected areas, Natura 2000 sites, the regional ecological network, and environmental, historical, and cultural assets (Fig. 1). Conceived as a network of networks (ecological, historical, and recreational), this multifunctional structure forms the basis for a territorial development project that "evolves in reticular forms" [12]. Planning thus adopts a dual approach: networked interpretation of natural systems and multifunctional ecological reticularity linked to cultural continuity. Its dynamic governance engages diverse actors to foster shared cultural identity and sustainable local development.

Currently undergoing revision, the PTR seeks alignment with the aforementioned strategies, reinforcing its role as a structural, strategic, and regulatory framework for territorial governance. A key innovation is the redefinition of the 33 Territorial Integration Areas (*Ambiti di Integrazione Territoriale*—AIT), which now serve as critical interfaces between regional, intermediate, and municipal planning levels. Each AIT is assigned specific objectives, allowing for targeted evaluation of planning interventions and prioritization of resource allocation. In addition to reorganizing its specific objectives around five strategic axes, the PTR introduces a new overarching goal: digitalizing and simplifying public administration processes.

2.2 Planning Protected Areas

Among Piedmont's territorial planning tools, park plans (*Piani d'Area*) for protected areas represent an excerpt of the PTR and must align with the PPR. As such, they are essential for implementing integrated planning at intermediate and supra-local

Fig. 1 The landscape connection network. *Source* Regional landscape plan of piedmont

scales, especially within territories managed by the Piedmont Region through dedicated management authorities (*Ente di Gestione delle Aree Protette*). To date, 13 of 22 protected areas have approved plans, with several other procedures underway in key areas of the Metropolitan City of Turin. In 2024, Piedmont issued strategic guidelines for developing new park plans, acknowledging their role not only in enhancing natural and cultural resources and biodiversity but also in advancing climate change adaptation strategies and improving ecosystem services provision. Since the 1970s, European policies have positioned protected areas as strategic actors in environmental and territorial planning [13]. Contemporary park planning must address global changes by assigning social, spatial, and temporal value to environmental policies that must be context-sensitive, differentiated upon resource diversity, and clearly communicated. This requires "the integration of knowledge, decisions, evaluations, and interventions" [14], and a shift from hierarchical governance toward a dialogical approach that aligns policies across scales and reinforces the central role of landscape, as emphasized in the PPR.

Planning and managing protected areas require a holistic vision integrating nature conservation, biodiversity protection, and development policies into forward-looking strategies. This is especially crucial given the scale and socio-territorial relevance of protected areas (24% of Europe and over 11.5% of Italy), particularly in relation to growing interactions with urban contexts [15]. In Piedmont, park plans, such as the Piedmontese Po park currently under development with our scientific coordination, are key tools for coordinating regulatory, local, and conservation actions, resolving tensions between environmental, landscape, and ecological issues, and aligning with the PPR's landscape enhancement goals. This pivotal role can, however, generate tensions within protected areas, especially between state authority over environmental and landscape protection and regional powers in territorial planning, which inevitably intersect. The Piedmontese Po park plan exemplifies a first attempt to integrate these competencies. As an operational element of the PTR, it acts as a primary land management tool, aligning landscape and biodiversity policies with those for sustainable development. Moreover, these plans aim to link protection and development strategies, anchoring parks within their broader territorial context. Rather than merely updating conservation policies, they should guide municipal governance, address local criticalities, and articulate a supra-local, long-term vision (Fig. 2). Crucially, they must be seen not as inter-municipal plans, but as strategic tools for confronting today's environmental and territorial challenges.

2.3 The Intermediate Level

Regional and provincial plans must be complemented by additional tools to articulate integrated biodiversity policies at the local level effectively. In Piedmont, a notable example is "The Guidelines for the Ecological Network", incorporated into the Provincial Coordination Plan of the former province of Turin. These guidelines define a strategic framework for creating the Provincial Ecological Network system to support land-use planning through biodiversity conservation. It contributes to limiting land consumption, enhancing and expanding ecosystem services, and promoting the sustainable use of natural resources and landscape assets.

This methodology is strategically valuable due to its simplicity in interpreting biodiversity from the intermediate to local scale. It enables the assessment of the ecological functionality of the territory by analyzing different land use types, based on Land Cover Piemonte, through five key indicators: naturalness, relevance for preservation, fragility, extroversion, and irreversibility. These indicators provide a comprehensive framework for evaluating the ecological status of areas. Integrating

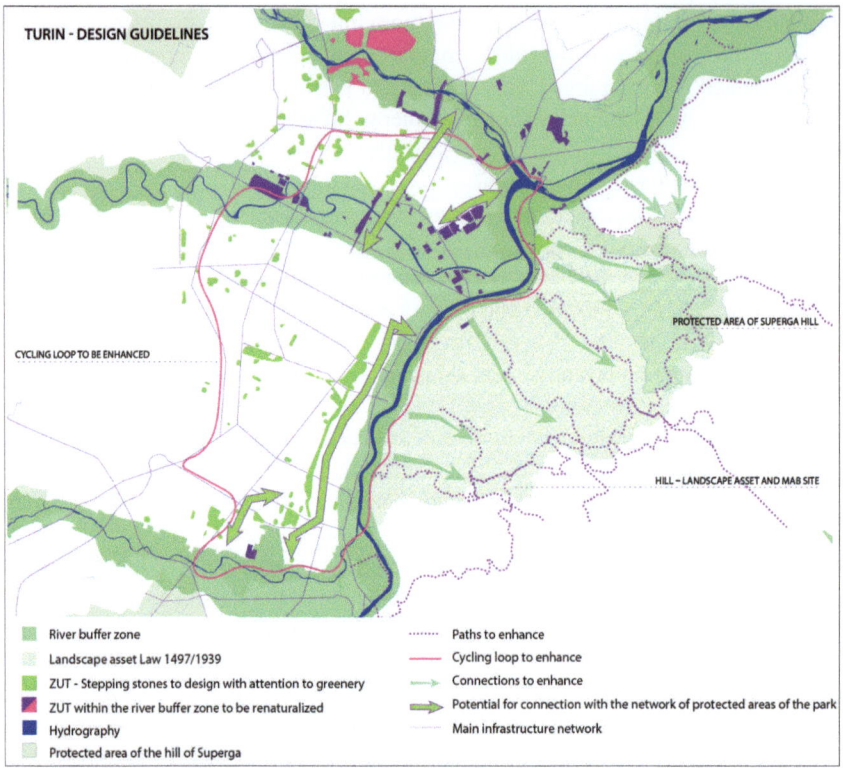

Fig. 2 Design guidelines for the city of Turin within the Piedmontese Po Park. *Source* Adapted from an image by Ardito

these results into a large-scale biodiversity map offers a clear visualization of the ecological structure of the area. This map identifies: (i) structural elements of the network (primary ecological reticularity), which include areas of high to moderate ecological functionality and zones hosting specific conservation values due to their high naturalness and biodiversity significance; (ii) the priority expansion areas, which are territories with residual ecological functionality where interventions are urgently needed to enhance the primary ecological network. These areas are subdivided into connection areas and buffer zones (within 50 m) adjacent to structural elements; potential expansion areas, i.e., areas where residual ecological functionality exists and where habitat and species conservation actions could be effectively implemented (Fig. 3).

By experimenting with this method, it is possible to assign ecological meaning and ecosystem value to various land uses. This approach goes beyond merely analyzing the state of naturalness and biodiversity at different scales; it enables the definition of priorities for conservation and enhancement actions across the territory. For example, in the structural scheme of the green infrastructure within the ongoing Metropolitan General Territorial Plan, the method allows for evaluating the ecological quality of

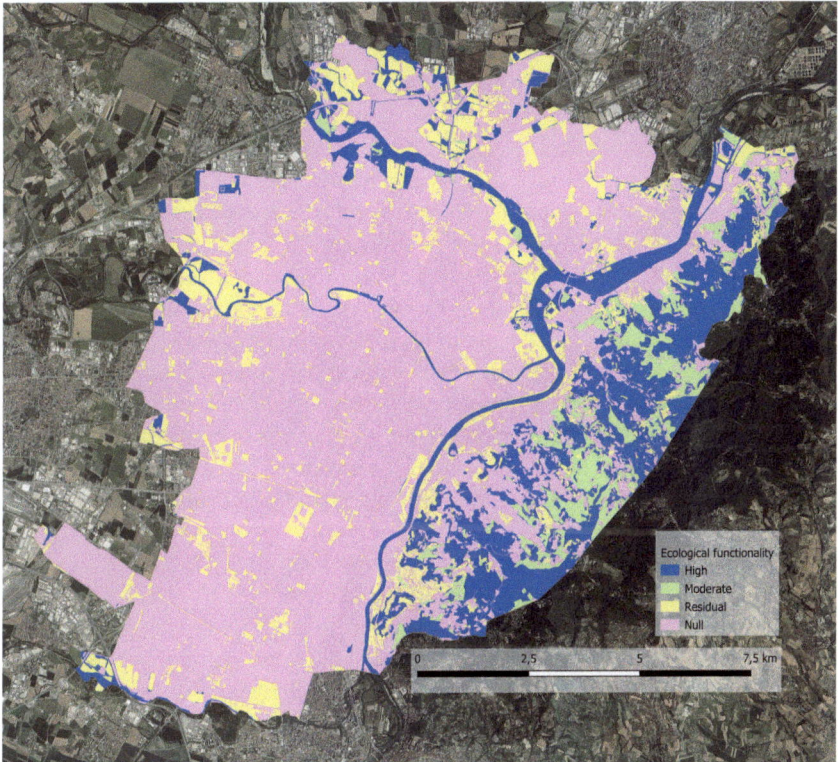

Fig. 3 Ecological functionality in the city of Turin. *Source* Elaboration by La Riccia

the territory and supporting the multiscale design of ecological networks, integrating the regional, provincial, and local levels.

3 Conclusions

The challenge of enhancing biodiversity within territorial and landscape policies, as emphasized by the Nature Restoration Law, introduces significant implications for spatial planning and urban design. It reflects a broader paradigmatic shift in our relationship with nature, echoed in the 2021 IUCN World Conservation Congress, under the banner "Our nature, our future", which promoted a nature-positive economy and nature-based recovery as central pillars of biodiversity action. These frameworks urge planners and policymakers to move beyond outdated models [16], advocating for an integrated approach that reconciles site-based conservation with sustainable development. Addressing the legacy of the so-called "great acceleration" in biodiversity loss requires embedding ecological transition across all planning scales through concrete, multi-level governance actions. In this context, distinguishing and connecting sustainability and resilience conceptual frameworks [17] becomes crucial in enabling a complex, trans-scalar planning approach. Such an approach expands the scope of territorial governance, demanding cross-sectoral collaboration among institutions, communities, and socio-economic stakeholders. It supports a broad agenda, from natural regeneration and biodiversity enhancement to climate neutrality, integrated risk management, and the broader energy transition.

In such contexts, park plans emerge as pivotal tools in this new planning ethos, particularly in regions like Piedmont with diverse landscapes (including mountain, rural, water, and peri-urban). Far from being limited to conservation instruments, they act as strategic tools that couple protection with innovation [18]. These "special territories" function as guardians of biological and cultural diversity and serve as antifragility reserves and testbeds for nature-based solutions, promoting sustainability, green economy initiatives, and human and ecosystem health. In this light, park plans contribute to a cultural shift that reimagines the relationship between humans and the Earth, aligning with global frameworks and national and regional directives.

Within this framework, area-based nature and biodiversity conservation strategies, such as protected areas, are increasingly seen not only as biodiversity reservoirs but also as catalysts for integrated territorial projects. By linking protected and surrounding areas through the lens of green and blue infrastructure, these strategies support ecological connectivity, enhance ecosystem services, and foster multifunctional landscapes. Urban and peri-urban protected areas and other green areas must be recognized as integral elements within a broader, trans-scalar ecological network.

Meeting these challenges requires planning and design approaches that embed biodiversity as a strategic value across all governance levels. Such integration advances both sustainability and resilience goals, while affirming the essential interdependence between people and nature. Ultimately, embracing the transformative alliance between protected areas, landscape, biodiversity, and territorial policies offers a proactive path to address the interconnected ecological, climatic, and societal crises of our time.

References

1. Dinerstein E, Vynne C, Sala E et al (2019) A global deal for nature: guiding principles, milestones, and targets. Sci Adv 5(4). https://doi.org/10.1126/sciadv.aaw2869
2. Gambino R, Peano A (eds) (2015) Nature policies and landscape policies. Towards an alliance. Springer, Cham
3. Schama S (1995) Paesaggio e memoria. Mondadori, Milano
4. Raffestin C (2015) From the territory to the landscape: the image as a tool for discovery. In: Gambino R, Peano A (eds) Nature policies and landscape policies. Towards an alliance. Springer, Dordrecht, pp 93–101. https://doi.org/10.1007/978-3-319-05410-0_10
5. APAT—Agenzia per la Protezione dell'Ambiente e per i Servizi Tecnici (2003) Gestione delle aree di collegamento ecologico funzionale. Indirizzi e modalità operative per l'adeguamento

degli strumenti di pianificazione del territorio in funzione della costruzione di reti ecologiche a scala locale. Manuali e linee guida 26/2003
6. Giudice B, Novarina G, Voghera A (eds) (2023) Green infrastructure. Planning strategies and environmental design. Springer, Cham
7. Mela A, Battaglini E, Palazzo AL (2024) La società e lo spazio. Quadri teorici, scenari e casi di studio, Carocci Editore, Roma
8. Voghera A (2015) Regional planning for linking parks and landscape: innovative issues. In: Gambino R, Peano A (eds) Nature policies and landscape policies. Towards an alliance. Springer, Dordrecht, pp 137–144. https://doi.org/10.1007/978-3-319-05410-0_14
9. La Riccia L (2015) Nature conservation in the urban landscape planning. In: Gambino R, Peano A (eds) Nature policies and landscape policies. Towards an alliance. Springer, Dordrecht, pp 157–164. https://doi.org/10.1007/978-3-319-05410-0_17
10. Giudice B, Voghera A (2024) Planning for landscape and heritage. A community perspective to overcome risks and vulnerabilities in the Italian case study. Int J Disaster Risk Reduct 110:104610. https://doi.org/10.1016/j.ijdrr.2024.104610
11. Colavitti AM, Serra S (2021) Regional landscape planning and local planning. Insights from the Italian context. J Settl Spat Plann 7:81–91
12. Terzuolo PG (2018) Ambiente e natura nel contesto del Piano paesaggistico regionale. In Cassatella C, Paludi G (eds) Il piano paesaggistico del Piemonte. Atti e Rassegna Tecnica LXXII-3:88–92
13. Gambino R (1996) Progetti per l'ambiente. Franco Angeli editore, Milano
14. Gambino R (1995) Separare quando necessario, integrare ovunque possibile. Urbanistica 104:57–65
15. Giudice B, Negrini G, Voghera A (2023) Il ruolo delle aree protette per la biodiversità urbana. Urbanistica Informazioni 308:15–19
16. Elhacham E, Ben-Uri L, Grozovski J, Bar-On YM, Milo R (2020) Global human-made mass exceeds all living biomass. Nature 588:442–444
17. Voghera A, Giudice B (2019) Evaluating and planning green infrastructure: a strategic perspective for sustainability and resilience. Sustainability 11(10):2726
18. Gambino R (1997) Conservare, innovare: paesaggio, ambiente, territorio. UTET, Torino

Open Access This chapter is licensed under the terms of the Creative Commons Attribution-NonCommercial-NoDerivatives 4.0 International License (http://creativecommons.org/licenses/by-nc-nd/4.0/), which permits any noncommercial use, sharing, distribution and reproduction in any medium or format, as long as you give appropriate credit to the original author(s) and the source, provide a link to the Creative Commons license and indicate if you modified the licensed material. You do not have permission under this license to share adapted material derived from this chapter or parts of it.

The images or other third party material in this chapter are included in the chapter's Creative Commons license, unless indicated otherwise in a credit line to the material. If material is not included in the chapter's Creative Commons license and your intended use is not permitted by statutory regulation or exceeds the permitted use, you will need to obtain permission directly from the copyright holder.

Cities Walk, Forests Run: Trees and Forests as Nature-based Solutions in Transforming Biocities

Fabio Salbitano, Giuseppe Scarascia Mugnozza, and Marco Marchetti

Keywords Trees · Forest · NbS · BioCities · Transition · Policy recommendations

1 Introduction

Urban environments cause biodiversity loss and are facing complex challenges, including rising temperatures, air and water pollution, social inequities. With a growing over half of the global population living in cities, the pressure on natural systems continues to intensify. The BioCity concept offers a regenerative paradigm: cities as life-supporting systems, inspired by the resilience, adaptability, and circularity of nature. This model urges a shift from extractive and linear development to integrated systems where green infrastructure (GI) plays a structural role. In this context, trees and forests are not merely ornamental but foundational components of BioCities. They serve as bio-infrastructure, providing vital ecosystem services, enhancing urban resilience, and reconnecting humans with natural cycles. By embedding trees and forests in adaptive spatial planning, cities can foster co-evolutionary relationships between society and ecosystems. The BioCities framework [1] integrates principles from landscape ecology, circular bioeconomy, and systems thinking.

F. Salbitano
University of Sassari and Fondazione Alberitalia, Sassari, Italy

G. Scarascia Mugnozza
European Forest Institute, Biocities Facility, Rome, Italy

M. Marchetti (✉)
Sapienza University and Fondazione Alberitalia, Rome, Italy
e-mail: ma.marchetti@uniroma1.it

© The Author(s) 2026
A. De Toni et al. (eds.), *Nature-Positive Cities: Adaptive Spatial Planning in Italy for an Ecological Urban Transition*,
PoliMI SpringerBriefs, https://doi.org/10.1007/978-3-032-06617-6_8

A BioCity operates as a social-ecological system with characteristics such as multifunctionality, connectivity, metabolic efficiency, and regenerative capacity. Within this system, forests and green infrastructures act as mediators between anthropogenic and natural processes. Urban forests are increasingly recognized for their carbon sequestration potential [2, 3] microclimate regulation, air purification, and psychological benefits. Their multifunctionality aligns with this vision: reducing emissions, enhancing biodiversity, improving public health, and enabling adaptive reuse of urban land. Forests and trees become pivotal in operationalizing BioCity principles across different spatial scales.

2 Planning BioCities

Nature-based solutions (NbS) are defined by the European Commission and IUCN as actions inspired and supported by nature that address societal challenges while delivering biodiversity and ecosystem services benefits. NbS encompasses a spectrum of interventions from urban forests to biowalls, green roofs, rain gardens, and riparian buffers. According to the NbS catalogues [4, 5] GI provides cross-cutting benefits for climate mitigation, disaster risk reduction, and social resilience. For example, tree-lined streets and green corridors mitigate urban heat islands and support biodiversity connectivity. Green roofs and walls enhance insulation and stormwater retention. Forest patches and peri-urban woodlands can store carbon and improve air quality while offering recreational and aesthetic value.

Planning BioCities requires therefore a paradigm shift in how urban space is conceptualized and governed. Adaptive spatial planning supports dynamic, iterative decision-making rooted in systems thinking and long-term ecological goals. We propose a five-phase framework:

- Visioning. Establishing shared goals for ecological health, biodiversity, and social equity.
- Assessment. Mapping urban metabolism, ecological baselines, ecosystem services supply and demand.
- Design. Co-producing spatial configurations for NbS deployment (e.g., forest belts, green corridors).
- Implementation. Mobilizing policy instruments, fiscal incentives, zoning tools, and community participation.
- Monitoring and Adaptation. Tracking ecosystem performance and feedback integration.

3 Mosaic of Forest-Driven Urban Transformation

In the unfolding era of planetary urbanism and urbanization, climate volatility and biosphere threats, cities face mounting challenges: extreme weather, ecological degradation, public health crises, rising social inequity. Against this backdrop, trees and urban forests are emerging not merely as amenities, but as strategic infrastructure—nature-based solutions (NbS) capable of reshaping urban governance. They are not passive background features, but dynamic agents in the transformation of cities into adaptive, resilient "biocities."

Rather than prescribing universal templates, case studies and practice-based evidences can reveal how local ecological, political, and cultural conditions shape the design and governance of urban forestry projects. Each example is a micro-laboratory, offering insights into what works, what doesn't, and why—especially when sustainability intersects with equity, technology, and community agency. Urban forestry's rising role stems from its multifunctionality. Trees moderate microclimates, sequester carbon, support biodiversity, filter air and water, and enhance mental and physical well-being. But these ecological services alone don't explain their transformative potential. What elevates trees and forests to the status of NbS is their capacity to bridge environmental function with civic engagement, territorial justice, and cross-sectoral governance. This paragraph curates a global spectrum of practices drawings from diverse evidence streams: satellite monitoring, health impact assessments, participatory governance models, and biodiversity metrics. The aim is not to celebrate "best practices" as static ideals, but to highlight replicable governance principles and adaptive design pathways. By focusing on "cities that move and forests that run," key themes explored in the the case studies include:

- Integrated Urban Planning: How trees are embedded not as afterthoughts but as drivers of land-use policy, transportation networks, and housing developments.
- Data-Driven Interventions: From urban canopy mapping to thermal risk indexes, cities are deploying sophisticated tools to target green interventions precisely where they are needed most.
- Governance Innovation: New institutions, such as Milan's *Forestami* or Guadalajara's *AMBU*, illustrate how multi-level and community-anchored models outperform siloed approaches.
- Equity and Justice: Whether through Toronto's energy poverty metrics or Medellín's job quotas, forests are being intentionally used to rebalance access and opportunity in divided urban landscapes.
- Cultural Adaptation: as shown in Kigali and São José dos Campos, grounded knowledge systems—from indigenous ecological governance to spiritual values attached to trees—are essential for context-sensitive design.

The purpose of following case studies is to generate a working theory of change: how cities can guide transition toward regenerative futures by embedding urban nature into the core of their adaptive planning. *Transformation of cities into biocities is already underway—and trees are not walking, they are running ahead.*

3.1 Trees, Forests and NbS as Indicators of Biocities Transformation

Across continents only the most mature transformations towards biocities, are recognizing that *the city is hosted by nature and not the opposite.* A new governance model is taking shape where trees and forests serve as operational tools for urban transformation. The case studies reveal three converging trajectories: ecological functionality, participatory governance, social equity.

In **Barcelona**, urban forest is treated as strategic infrastructure. The Green Infrastructure and Biodiversity Plan integrates 783 ha of green space and 49 km of ecological corridors. Data-driven tools like the Urban GI Index guide targeted interventions, particularly in heat-vulnerable districts. Cooling interventions in schools and greenways in underserved areas, bring to reduced cardiovascular hospitalizations, illustrating the health co-benefits of this ecological planning.

Medellín (Colombia), once known for its degraded urbanity, has emerged as a global exemplar in Biocity transformation by implementing the innovative "Green Corridors" infrastructure, converting 18 major roads and 12 waterways into interconnected blu/green spaces that serve both ecological and social functions. They significantly mitigate urban heat island effect: between 2016 and 2019, these areas experienced a reduction in average air temperatures from 31.6 to 28.1 °C, and surface temperatures dropped by over 10 °C (the 2019 Ashden Award for Cooling by Nature). The effect enhances thermal comfort and reduces energy consumption, and have improved air quality, by reducing PM2.5 levels by 1.55 $\mu g/m^3$. The introduction of over 880,000 trees and 2.5 million plants has created habitats for various species, bolstering urban biodiversity. Moreover, the city's botanical gardens have trained disadvantaged people to become urban gardeners and planting technicians, providing them with employment opportunities and fostering community stewardship. Medellín's approach includes participatory budgeting. Inclusive governance model ensures that the development and maintenance of green spaces reflect the community's needs and values. This holistic approach serves as a blueprint for cities worldwide aiming to transition into biocities that prioritize both environmental health and social well-being.

In **Kigali**, Rwanda, climate adaptation is paired with gender-responsive governance. SUNCASA Project exemplifies how urban forestry can serve as a powerful NbS when embedded within a holistic framework of climate adaptation, social equity, and gender empowerment. The initiative targets degraded urban and peri-urban watersheds with a comprehensive reforestation strategy designed to mitigate flood and landslide risks while enhancing local livelihoods. Between 2023 and 2026, 2.3 million indigenous trees have been planting, including 125.000 fruit-bearing species intended to support agroforestry-based income generation. This ecological and economic focus underscores the project's alignment with sustainable development goals, enhancing food security, reducing vulnerability to climate shocks, with a strong gender-transformative design. Women constitute 50% of participants in seedling production and planting activities. These women-led cooperatives are not only responsible for nursery operations but also manage key segments of the seedling supply chain, reinforcing female agency in green economy development [6]. The project has established 395 ha of gully stabilization buffers, strategically placed to restore hydrological integrity and intercept runoff, to reduce flood risk for approximately 975,000 residents and landslide occurrence (up to 42%, [7]), particularly in informal low-lying neighborhoods highly susceptible to climate-induced hazards. As cities in the Global South, Kigali's approach demonstrates the scalable potential of inclusive, NbS adaptation strategies.

Melbourne stands at the forefront of urban climate adaptation, leveraging its Urban Forest Strategy to transform GI into a dynamic tool for resilience. Confronted with escalating heatwaves, prolonged droughts, and urban densification, the city has implemented a multifaceted approach that integrates species diversification, water-sensitive urban design, and digital monitoring systems. The Urban Forest Diversity Guidelines deliberate shift towards thermally resilient tree species and advocate for a diversified urban canopy (*Eucalyptus polyanthemos* and *Brachychiton populneus* have been prioritized for their adaptability to increased temperatures and reduced water availability). Melbourne has also embraced water-sensitive urban design (WSUD) principles. Innovative infrastructure, including tree pits with subsurface infiltration systems, has been implemented to enhance soil moisture retention and reduce surface runoff, supporting tree health during periods of water scarcity. Interventions can reduce runoff volumes by approximately 33%, contributing to both urban cooling and water conservation. for. The deployment of real-time monitoring of tree health through sensors (Internet of Things—IoT) enables continuous assessment of vital parameters such as soil moisture, temperature, and tree growth rates. This data-driven approach facilitates timely interventions and adaptive management, ensuring that the urban forest remains robust amidst changing climatic conditions. Preliminary data indicates a 92% survival rate for newly planted trees, even during extreme heat events reaching 40 °C. This integrative model offers valuable insights for cities worldwide seeking to fortify their urban ecosystems against the multifaceted impacts of climate change.

In **Milan**, the Bosco Verticale reimagines vertical urban space as biodiverse habitat. This architectural icon grew back-to-back to a wider reforestation campaign - *Forestami*- targeting peri-urban floodplains for biodiversity gains and microclimate regulation. It exemplifies how designing innovation nature-inspired can be an attractive testimonial for activating green policies and initiatives while scaling into metropolitan ecological planning [8].

Singapore presents the hyper-dense application of NbS. Aerial canopy bridges, mangrove buffer zones, biome-specific reforestation protect biodiversity while generating millions in ecosystem service value, including climate resilience and tourism.

São José dos Campos, Brazil, offers a bottom-up model where residents flag tree-planting sites via a public database. QR-coded trees educate citizens, and planning regulations enforce urban greening as a civic duty. This approach exemplifies democratic ecological governance.

Curitiba, Brazil continues to lead with its integration of urban agriculture and GI. Programs like Urban Vegetable Gardens and Honey Gardens provide food security and social inclusion, while large-scale tree planting reinforces the city's commitment to ecological justice and community resilience.

Shenzhen, China, one of the fastest-urbanizing regions, has implemented a "Belt and Cluster" plan—infusing greenways and sponge city infrastructure to manage flood risk, support biodiversity, and buffer development. Urban forests are climate tools and status symbols of a new ecological urban identity.

Copenhagen merges dense planning with deliberate ecological zoning. The Urban Nature Strategy (2015–2025) and "Finger Plan" ensure that urban expansion is interlaced with protected green wedges, combining urban resilience with biodiversity corridors. Citizen participation is central to implementation.

Berlin has embraced a greening long-term strategy for *"Grünberlin"* through initiatives like the Green Moabit project and the city's Biodiversity Strategy. These tie into city's broader ecological governance model that values participatory planning and adaptive reuse of vacant urban spaces.

In **Padua, Pesaro,** and **Bologna** (Italy), local administrations have embraced NbS as critical to sustainability planning. Padua's mapping of public and private green space supports equity-focused zoning decisions. Bologna's transformation of brownfields into forests (e.g., Prati di Caprara) shows how forest infrastructure can reclaim forgotten spaces as ecological commons. Pesaro FEVER Strategy (Ecological functions of Urban green spaces) bridge the management of trees, urban forests and urban parks to the overall town planning integrating offices as Environment, School and Education, Culture, Public Works, Civil protection and security.

Rabat, Morocco, anchors NbS in national climate policy, linking biodiversity, wetland restoration, and urban parks with its Nationally Determined Contributions. Projects like these show how international frameworks like the UNFCCC can drive local greening.

Many other cities (e.g., **Turin, Rome, Paris, Ljubljana,** and **Ulaanbaatar**) are testing governance innovations, from legal mandates for GI to participatory reforestation schemes. Many are embedding urban forestry in climate action plans, zoning frameworks, and funding strategies. Finally, programs like Tree Cities of the World, promoted by the Arbor day Foundation and FAO, active in cities from **New York** (among the largest) to **Lignano Sabbiadoro** (Italy, among the smallest), build global momentum by tending to harmonize and standardize the supporting criteria for a proactive urban forest governance. Metrics, public engagement, and shared learning are helping local governments frame trees as core infrastructure. As forests "run" ahead of cities—regenerating degraded land, shaping urban form, and reframing public space—the challenge now lies in governance: how to embed these living systems into institutions, budgets, and culture. The future Biocity is not built around concrete, but around care—for people, place, and planet.

3.2 Comparative Insights

A mosaic of forest-based interventions reveals diverse, locally anchored pathways toward biocity transformation. From watershed-based agroforestry to digitally monitored tree resilience, cities across continents are operationalizing NbS in uniquely adaptive ways. Beneath this diversity, some comparative insights emerge (Table 1).

Governance matters more than design. Cities with integrated, cross-sectoral governance models outperform those with siloed environmental departments. Institutional alignment is critical for scale and longevity.

Equity is not a side effect, but a core metric. Gender-quota nurseries and energy-poverty forest expansion explicitly center justice in ecological planning. The success of NbS must be judged not only by environmental impact, but by distributive fairness.

Ecological Intelligence Enhances Adaptation

Cultural framing shapes success. Cultural narratives—from Indigenous stewardship to civic gamification—drive community engagement and long-term stewardship of green infrastructure.

Multi-scalar design unlocks systemic benefits. Cities that integrate NbS across scalesgenerate systemic co-benefits such as flood reduction, urban cooling, and mental health improvements.

Data drives decisions, but needs interpretation. Cities must pair data with participatory governance to ensure community-aligned action.

Replication requires adaptation, not copying. The case studies reveal that NbS success is context-dependent. Exporting models must be matched with ecological, social, and institutional tailoring. These insights suggest a planetary palette of possibilities, where forests become adaptive agents, not static installations—running ahead of cities in function and imagination.

Table 1 Comparative insights according to governance model clusters

City	Country	Governance model	Participation	NbS focus	Climate benefit
Phoenix	USA	Community co-governance	Workshops	Tree equity, heat reduction	Heat mortality reduction
Medellin	Colombia	Community co-management	High	Cooling, biodiversity, jobs	2.8 °C UHI reduction
Toronto	Canada	Indigenous co-governance	cultural integration	Carbon justice, Indigenous knowledge	8000t CO_2/yr
Cape Town	RSA	Local/NGO partnerships	Strong	Wetlands, education	Water retention
Berlin	Germany	Participatory	Moderate	Brownfield regeneration	Habitat restoration
Montpellier	France	Public engagement	Active	Environmental education	Urban reuse
Barcelona	Spain	Multi-level	Co-design	Heat, biodiversity, health equity	4 °C school cooling
Melbourne	Australia	District/Neighbourhood	consultative	Drought, heat resilience	33% runoff reduction
Curitiba	Brazil	Municipal	High	Food security, greening	Urban cooling
Pesaro	Italy	Municipal and regional	Supported	Sustainable urbanism	Urban cooling
Bologna	Italy	Municipal and regional	Strong	Urban forest regeneration	Heat island mitigation
Padua	Italy	Municipal and regional	Active	Spatial equity	Heat mitigation
Milan	Italy	Municipal, Public–private	Moderate	Microclimate, biodiversity	7 °C surface cooling
Copenhagen	Denmark	Regional masterplan	High	Biodiversity corridors	Urban cooling
Rabat	Morocco	National integration	Moderate	Climate policy	Urban resilience
Granada	Spain	Municipal Spatial planning	Active	Heat risk mitigation	Heat reduction
Shenzhen	China	State-led	Emerging	Greenways, sponge city	Flood control

(continued)

Table 1 (continued)

City	Country	Governance model	Participation	NbS focus	Climate benefit
Singapore	Singapore	State-led with incentives	Moderate	Biodiversity, climate resilience	$70 M/year ecosystem value
Kigali	Rwanda	Transnational partnership	High	Watershed restoration, gender equity	15% flood risk reduction

4 Technical Considerations for Forest NbS in Urban Planning

Implementing NbS grounded in urban forestry is technically complex and highly context-sensitive. While the concept of "planting trees" appears simple, the design, integration, and long-term management of forest-based NbS requires rigorous interdisciplinary coordination, evidence-based methods, and a systems-thinking mindset. Key technical considerations include (Fig. 1):

- Urban environments are characterized by compacted soils, air pollution, altered hydrological cycles, and urban heat islands. **Selected species** must therefore be thermally resilient, tolerating extreme heat and irregular precipitation. Non-invasive and site-adapted, to preserve native ecologies and reduce maintenance costs.

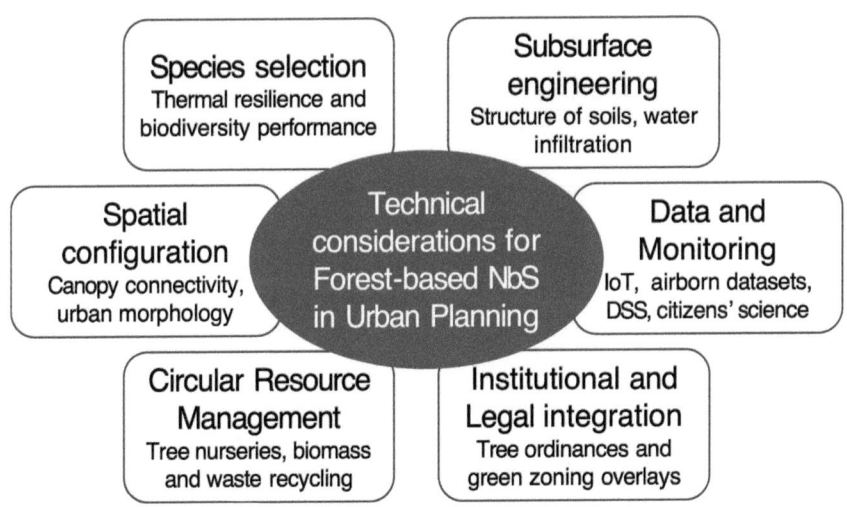

Fig. 1 Key technical considerations for the integration of forest-based nature-based solutions (NbS) into urban planning

- **Spatial Configuration and Urban Morphology** require strategic placement—along streets, in parks, on rooftops, and within riverine systems— since trees can: (i) increase canopy connectivity and urban ventilation, (ii) improve thermal regulation, (iii) reducing localized UHI effects, (iv) enhance rainwater absorption and prevent runoff and flooding.
- **Urban soils** are often degraded and compacted, requiring bio-engineered solutions. Enhanced substrates and mulch systems improve root development. **Subsurface infiltration** pits and biowalls support stormwater capture, improving tree health and reducing flood risks. Gully stabilization restores eroded landscapes while acting as flood buffers. Soil microbiology is increasingly recognized as a determinant of NbS longevity, suggesting a need for mycorrhizal assessments and microbial inoculation in urban soils.
- **Monitoring, Data, and Digital Infrastructure**. NbS are living systems—dynamic and sensitive. Thus, monitoring is critical: IoT-based sensors provide real-time data on soil moisture, growth rates, and stress signals. Remote sensing and LiDAR enable canopy mapping, evapotranspiration modeling, and shadow cast analyses. Decision support platforms (e.g., i-Tree, Green View Index) allow scenario planning and value quantification for carbon, cooling, and pollution mitigation. These tools support adaptive governance—helping planners revise interventions based on real-time ecosystem performance.
- **Circular Resource Management**: a climate-aligned NbS strategy cannot ignore the resource lifecycle and the related environmental impacts generated from implemented initiatives and actions. For example: (i) trees must be sourced from tree nurseries following low-carbon, pesticide-free practices, (ii) urban wood waste can be reused for construction or energy, supporting bioeconomy models, (iii) tree death or replacement should be managed with composting, mulching, or biochar generation, (iv) LCA help to measure the embedded carbon and environmental impact of planting operations.
- **Institutional and Legal Integration** must consider tree ordinances, development codes, and green zoning overlays. Planning laws should require compensatory planting and GI credits for developers. Institutional arrangements enable coordination across different departments.

5 Challenges and Opportunities

The transition to BioCities demands rethinking the spatial, ecological, and governance logics of urban systems. Trees and forests, when embedded through adaptive spatial planning, are catalysts of regeneration and resilience for **key challenges and key opportunities**, such as:

- Governance Fragmentation: Forest-based NbS require interdepartmental collaboration—across environment, health, infrastructure, and urban planning. Yet, most cities operate in silos. Without dedicated institutions or coordinators, NbS become disjointed, underfunded, or symbolic.
- Land Access and Spatial Constraints: As densification increases, available land for large-scale green interventions shrinks. The challenge is finding creative integration points—on rooftops, walls, brownfields, under elevated roads—while resisting green gentrification.
- Monitoring Gaps and Evaluation Inconsistencies: Impact measurement remains inconsistent. Many cities lack of a initial set of data as: (i) Baseline ecological data, (ii) Longitudinal monitoring of benefits (e.g., carbon uptake, cooling), (iii) Social metrics (e.g., access equity, public health impacts). Without shared indicators, comparing and scaling NbS becomes difficult.
- Risk of "Greenwashing": cities risk presenting minimal greening as transformational. Without long-term maintenance plans, community involvement and ecological coherence, trees can become liabilities—not solutions.
- Equity Backlash and Gentrification: urban greening can inadvertently accelerate displacement. Neighborhoods made "greener" often experience increased property values and living costs, pricing out original residents—contradicting the inclusive goals of NbS.

– Policy Convergence: climate adaptation, public health, biodiversity, and green jobs now share overlapping goals. NbS offer a convergence policy tool that delivers cross-sectoral outcomes, qualifying for diverse funding streams.
– Decentralized and Participatory Models of green management are always successful—and community co-ownership of forests fosters care, equity, and democratic resilience.
– Digital Tools for Targeting and Inclusion. AI-powered equity maps, remote sensing tools, and open participatory platforms (e.g., urban forest dashboards) could allow precision targeting of vulnerable neighborhoods and transparent communication.
– Emerging carbon credit markets for urban greening, and ESG-linked investment funds, present new co-financing opportunities. Cities can frame urban forests as carbon-positive assets with measurable returns.
– Workforce and Skills Development: NbS can be a springboard for green jobs—from nursery operators to urban ecologists, arborists to community stewards, aligned with gender equity and youth employment agendas.

6 Conclusions and Recommendations

The transition toward BioCities is more than a visionary exercise; it is a necessary response to converging environmental, social, and climatic crises. Urban forests and tree-based NbS and GI offer a tangible, scalable, and inclusive strategy to reshape urban futures. This chapter has explored their role through the lens of adaptive planning, supported by global case studies, technical planning considerations, and governance insights [9]. Urban forests should no longer be regarded as marginal amenities but as fundamental components of urban infrastructure. Their capacity to deliver multiple ecosystem services—climate regulation, air purification, stormwater management, biodiversity support, and social cohesion—positions them as essential assets in achieving nature-positive cities. However, their effective deployment hinges on a deep reconceptualization of urban planning, one that embraces complexity, fosters inclusivity, and prioritizes long-term ecological resilience. To embed forest-based solutions within urban transitions, some key recommendations emerge.

(a) Integrate urban forestry into climate adaptation and mitigation strategies by aligning canopy goals with heat reduction targets, carbon neutrality plans, and environmental justice objectives.
(b) Institutionalize cross-sectoral governance frameworks that bridge environmental, health, housing, infrastructure, and education sectors to coordinate NbS efforts.
(c) Mainstream equity in spatial planning and funding by prioritizing interventions in underserved areas, embedding co-design processes, and ensuring sustained investment in tree stewardship and community engagement.
(d) Advance monitoring and evaluation systems using innovative technologies, citizen science, and ecosystem service accounting to track NbS performance, adaptively manage forests, and report transparently.
(e) Foster innovation ecosystems that connect urban forestry with digital tools, green job creation, and circular bioeconomy strategies.
(f) Anchor urban forest policies in legal frameworks through tree ordinances, biodiversity zoning codes, and development incentives. Regulatory clarity is essential for enforcement, compliance, and accountability.
(g) Invest in education, capacity-building, and green culture to cultivate long-term stewardship. Integrate urban forestry into school curricula, university programs, and professional training for planners, designers, and arborists.

The future BioCity is not only greener but more intelligent, just, and regenerative. In its vision, trees are not passive decorations but dynamic agents of change, anchoring urban metabolism in living systems. Planning with forests is planning with time, with care, and with the recognition that cities, like ecosystems, must evolve in symbiosis with the planet's limits.

References

1. Scarascia-Mugnozza GE et al (2023) Transforming BioCities. Springer
2. Nowak DJ et al (2013) Carbon storage and sequestration by urban trees in the USA. Environ Pollut 178:229–236

3. Fares S et al (2023) Mitigation and adaptation for climate change: the role of BioCities and nature-based solutions. In: Transforming BioCities
4. World Bank (2021) A catalogue of nature-based solutions for urban resilience
5. Morello E et al (2019) Catalogue of nature-based solutions for urban regeneration. Politecnico di Milano
6. Bidibura W et al (2023) Nature-based solutions 3:100055
7. Mugume I et al (2021) Int J Disaster Risk Reduction 56:102124
8. Forestami (2023) Annual impact report
9. Salbitano et al (2021) Forest policy and economics 133:102618

Open Access This chapter is licensed under the terms of the Creative Commons Attribution-NonCommercial-NoDerivatives 4.0 International License (http://creativecommons.org/licenses/by-nc-nd/4.0/), which permits any noncommercial use, sharing, distribution and reproduction in any medium or format, as long as you give appropriate credit to the original author(s) and the source, provide a link to the Creative Commons license and indicate if you modified the licensed material. You do not have permission under this license to share adapted material derived from this chapter or parts of it.

The images or other third party material in this chapter are included in the chapter's Creative Commons license, unless indicated otherwise in a credit line to the material. If material is not included in the chapter's Creative Commons license and your intended use is not permitted by statutory regulation or exceeds the permitted use, you will need to obtain permission directly from the copyright holder.

Nature-Positive: Transforming Cities and Landscapes with Scalable Strategies and Projects. Insights from LAND's Case Studies

Andreas Otto Kipar, Valentina Galiulo, Gloria Signorini, and Daniele Galimberti

Keywords Nature-positive · Climate resilience · Ecosystem services · Nature-based solutions · Stakeholder engagement · Sustainable development

1 The International and Local Geo-Political Context for the Development of Nature-Positive Strategies

As the degradation of ecosystems continues at an unprecedented rate, cities around the world face a critical challenge: how to grow while simultaneously regenerating the natural systems that sustain them. **The concept of "Nature-Positive" urbanism**—defined as development that halts and reverses biodiversity loss—has become a defining principle for ecological planning. Yet the capacity to implement these strategies depends heavily on both international frameworks and local geopolitical realities.

At the global scale, the Kunming-Montreal Global Biodiversity Framework (GBF), adopted in 2022 by 196 governments, marks a paradigm shift in planetary

---

A. O. Kipar · V. Galiulo · G. Signorini (✉) · D. Galimberti
LAND Srl, Milan, Italy
e-mail: gloria.signorini@landsrl.com

A. O. Kipar
e-mail: andreas.kipar@landsrl.com

V. Galiulo
e-mail: valentina.galiulo@landsrl.com

D. Galimberti
e-mail: daniele.galimberti@landsrl.com

governance. Unlike previous agreements, it includes quantified targets to be achieved by 2030 and clear links between biodiversity, economy, and human health. Target 12, for instance, calls for "urban and other areas to enhance biodiversity and green spaces" while ensuring equitable access to the benefits of Nature [1]. This signals an urgent invitation for cities to become active contributors to planetary restoration, not just passive recipients of its consequences.

Complementing this is the Taskforce on Nature-related Financial Disclosures (TNFD), which, like the Task Force on Climate-related Financial Disclosures (TCFD), provides a framework for identifying Nature-related dependencies, impacts, risks, and opportunities. Its integration into corporate and public planning creates a bridge between finance, governance, and ecological value [2]. The Science Based Targets Network (SBTN) similarly supports cities and businesses in translating global goals into measurable local targets through science-based methodologies [3].

However, while multilateral guidance provides direction, national frameworks often determine the feasibility of city-level action. For example, the UK's biodiversity net gain mandate requires developers to deliver at least 10% more biodiversity post-development compared to pre-development baselines—a legal obligation that provides cities with both a tool and a mandate [4]. Germany's Federal Nature Conservation Act includes biotope protection regulations that directly shape urban planning decisions, while Italy's National Strategy for Biodiversity 2030, aligned with the GBF, remains limited in its urban-specific provisions, creating challenges for city administrations attempting to lead on Nature [5].

In federal states or countries with strong local autonomy municipalities are often the pioneers of innovation, but are also left without sufficient legal and financial support. This imbalance creates a "policy vacuum," where cities are eager but underequipped to implement Nature-Positive policies [6]. In such cases, public–private partnerships and municipal alliances are essential, with cities like Milan and Hamburg demonstrating how decentralized action can bridge national shortcomings through innovation, stakeholder engagement, and cross-sector collaboration [7].

Local political will and civic culture play an equally decisive role. Cities vary significantly in their institutional capacity to develop and execute biodiversity strategies. A city like Durban, one of the top 5 Nature-Positive Cities for World Economic Forum (WEF), for instance, has embedded Nature stewardship into its planning identity, leveraging community participation and robust spatial planning tools [8]. In contrast, many urban governments face structural limitations that hinder long-term investment in ecological infrastructure, such as parks, green corridors, and restored wetlands. Political cycles, fiscal constraints, and competing urban demands often lead to the marginalization of Nature in planning processes.

Yet amid these challenges, a new form of action planning is emerging. Cities are no longer passive recipients of national mandates but have become active agents in the global sustainability discourse. Through networks such as C40 Cities, ICLEI's CitiesWithNature, and the CDP global disclosure system, local governments are sharing best practices, setting voluntary biodiversity targets, and shaping norms around ecological restoration [9]. These collaborations function as soft-power alliances that transcend traditional geopolitical boundaries, redefining leadership in the climate and biodiversity arenas.

Nature-Positive strategies also intersect with broader political dynamics, including climate migration, resource security, and urban inequality. As biodiversity loss intensifies urban vulnerability to extreme weather events, food system disruptions, and economic instability, **the integration of Nature into urban resilience strategies becomes not only an ecological necessity but a socio-political imperative**. The IPBES Global Assessment warns that over one million species face extinction, many within decades, directly threatening the ecological functions upon which cities depend [10].

In this context, Landscape Architecture and Ecological Urbanism emerge as more than professional practices—they are civic tools for navigating and negotiating the relationship between Nature and governance. Planners, designers, and policymakers are called to act as interpreters of both ecological data and geopolitical narratives, shaping interventions that are technically sound, socially just, and politically feasible.

2 LAND's Commitment to Ecological Transition: Addressing Nature-Positive Framework into Actions

By embracing the Nature-Positive framework within a robust understanding of geopolitical realities across global mandates, national legislation, and local dynamics, cities can lead a new era of urbanization where regeneration becomes the default, not the exception. The future of our urban ecosystems lies in the hands of those willing to see Nature not as a constraint to development, but as its precondition.

Drawing on direct experience within the WEF Taskforce for Nature-Positive Cities [11], **LAND has embraced this Nature-Positive mission in daily work, encouraging both companies and urban administrations to move beyond sustainability conceived as mere damage control**. The international landscape consultancy firm leverages the scientific expertise of LAND Research Lab srl (the think tank of LAND) to advance a regenerative model aimed at delivering Net-Positive outcomes for biodiversity and ecosystem health. Through Natural Capital Accounting (NCA), useful for quantifying ecosystem services such as temperature regulation, air and water quality, and resilience to extreme weather events, it also advocates for public–private collaboration to mobilize collective investment. This approach involves engaging all stakeholders in actions that enhance private sector productivity while also serving the public interest, including objectives such as public health, social resilience, and environmental well-being.

With the aim of making the principles of Nature-Positive Cities easily applicable, extending them to landscape design, LAND co-created with Porsche Consulting the "Nature-Factory Manifesto" [12]: a practical operational guide for integrating NCA into the daily operations of both public and private stakeholders, conceived as active agents of Nature-Positivity.

To turn policy into action by translating the complexity of frameworks into projects grounded in local contexts, the landscape consulting firm promotes a new understanding of the environment and public space, leveraging the measurable potential of sustainability to foster widespread urban and economic development. Working across continents compels to pursue global goals while crafting local solutions deeply rooted in the communities that inhabit each landscape.

Drawing on 35 years of expertise in design, advocacy, and policymaking, LAND offers a comprehensive perspective on frameworks, tools, and policies that address the concept of Natural Capital, aligning and interpreting these principles to transform them into actionable strategies, embedded within the life and development of cities.

As mediators between public and private sectors, Nature and architecture, and as bridge-builders across continents, it is fundamental to translate this vast global effort into streamlined workflows, clearer data sets, and practical solutions.

The next section highlights three LAND projects where Nature-Positive principles have been applied in both a cross-cutting and site-specific way, shaping the design of public spaces across diverse urban contexts. These include: the regeneration of a post-industrial landscape (Krupp Park, Essen); the integration of Nature into an arid climate context to address climate change challenges (Al Urubah Park, Riyadh); and a data-driven, community-engaged process of urban regeneration and transformation (T-Factor Bioscopium Project, Milan).

3 Nature-Positive Applications into International Landscape Projects

3.1 Krupp Park "Five Hills" in Essen (Germany): From Core to Margins, A New Quality of Life

Like other former industrial cities such as London, Barcelona, Milan, and Turin, the city of Essen has undertaken enormous efforts to restructure its industrial heritage and is today internationally recognized for pioneering green transformation.

Starting from the **Masterplan "Freiraum schafft Stadtraum" ("Open Space Creates City Space", 2005–2022**), a comprehensive urban development initiative developed by LAND in collaboration with the city of Essen. Centered on reconnecting and enhancing green and open space axes, the plan uses inner-city waterways as "rays" along which public spaces are opened up and visually connected. This approach not only improves spatial connectivity but also raises public awareness through strategic "staging" of these areas. Complementing this, the action program "ESSEN. New Ways to the Water" [13] focuses on linking green areas between the Emscher and Ruhr Valleys. LAND's role in both planning and advising contributed to Essen being named the European Green Capital in 2017. By 2021, numerous projects had been realized, following a five-phase model. The first four phases were

completed over 15 years, while the fifth compiles outcomes from 2006 to 2021 and sets a strategic direction for the program's next phase.

At the heart of this evolution lies Krupp Park—a 20 ha public landscape that is exemplary for developing large green–blue infrastructures through the revitalization of inner-city brownfield sites of the Ruhr metropolis. The park forms the green heart of the larger Krupp Belt redevelopment area, reconnecting over 230 ha of urban fabric that had long been inaccessible to the public. Once enclosed for nearly two centuries, this territory has been transformed into a generous and open landscape, where Nature, history, and contemporary urban life converge.

Divided into a northern and southern part, Krupp-Park has a combined area of 220,000 m^2 in total, marking the beginning of the urban development of the green Krupp Belt, integrating and linking up the ThyssenKrupp-Headquarters and the new Krupp-Boulevard to the existing environment. Adjacent to the interregional bicycle lane and pedestrian paths which develop along the former "Rheinische Bahn" railway track, a green link to the Ruhr-Area with high potential of touristic attraction was created.

Being the fourth largest local recreation area in Essen's urban space, the Park not only serves as a hotspot for multifunctional open-air activities, featuring diverse kinds of sports facilities as well as areas to host large cultural events. Krupp Park also fosters social cohesion in a densely built environment while restoring ecological balance and mitigating climate impacts. The design by LAND goes beyond aesthetics, integrating a diverse topography of meadows, reforested areas, and open clearings that reflect the historical landscape while supporting local biodiversity. 400,000 m^3 of excavated soil was reused in the park so that its characteristic five hills could be created despite the soil contamination from the former industry.

A major reforestation effort saw the planting of more than 11,500 trees, reviving 5.5 of the 8 ha designated for woodland and creating cooler microclimates crucial for vulnerable urban populations. Drawing from ecological systems thinking, the park incorporates an intelligent rainwater management system. All rainwater, including runoff from nearby rooftops, is collected, filtered, and returned to the environment via a 10,000 m^2 lake system—an emblem of circular water usage and climate-sensitive design. Only one-third of the park's surface is sealed.

Through its layered, Nature-based design, Krupp Park demonstrates how landscape architecture can re-stitch the urban fabric and activate disused industrial sites, providing an **adaptive green infrastructure** that addresses pressing urban challenges—heat stress, biodiversity loss, and social fragmentation (Fig. 1).

Fig. 1 View of Krupp Park "Five Hills" (Essen, Germany). Image from *Krupp Park "Five Hills"— From the middle to the edges: new quality of life*, Landscape Biennial, COAC and UPC. *Source* lan dscape.coac.net

3.2 Al Urubah Park in Riyadh (Saudi Arabia): Landscape Architecture Under Extreme Conditions

Exploring the unique challenges and opportunities presented by environments with naturally harsh climates allows us to shape landscape projects that push the boundaries of traditional design. With its rich cultural heritage, diverse ecosystems, and ambitious vision for the future, Saudi Arabia offers fertile ground for innovation. A fundamental shift in urban planning is currently underway here, echoing the Volkspark movement of the 1920s, which emerged from the need to address urban issues and the impacts of industrialization. In these contexts, investing in green spaces as an act of collective survival and environmental justice means redefining the role of public space as both ecological and social infrastructure.

It is within this framework that Al Urubah Park takes shape, MENA region's first digitally driven park and a flagship project within Saudi Arabia's Green Riyadh Program [14], developed through constant collaboration with the Royal Commission for Riyadh City (RCRC). The existing natural network of the City becomes the main framework to activate the above-mentioned strategy and contributes to creating **extensive open green spaces in the city as a sustainable tool to fight**

Fig. 2 Rendering of Al Urubah Park Water Plaza—South Promenade. Image from LAND, *Al Urubah Park. Source* landsrl.com

climate change, capable of generating functional benefits in terms of environmental protection.

The park rises on a former military training ground in the eastern side of the capital, with strategic continuity 3 km far from King Salman Park, which is set to become the world's largest urban park. Spanning 75 ha, Al Urubah Park seamlessly integrates into Riyadh's expanding green infrastructure (Fig. 2).

Its design is based on "Nature-Positive" principles, with sustainable water management features such as drainage wells and trench channels supporting a 475,000 m^3 stormwater detention basin, enhancing both resilience and local biodiversity. The park also includes 40 ha of native vegetation, including indigenous trees and shrubs along drainage routes, and multi-species associations, which help combat the urban heat island effect, improve air quality, and provide habitat for local wildlife. Thus, the Inner Park's layout ensures total soil permeability, with no underground infrastructure, guaranteeing 100% permeability and supporting the sustainable water management strategy.

In this unique scenario, the digital landscape, envisioned as a multimedia extension of the natural landscape, becomes an integral part of visitors' experience, thanks to a multitude of devices with augmented reality functions. Technology serves not only to celebrate art, Nature, culture and traditions but also to engage the public and heighten awareness of environmental challenges.

Aligned with Saudi Vision 2030 and its Quality-of-Life Program, the park will serve as a community-centered space with diverse recreational areas, including play zones, sports facilities, and a 2.7 km Garden Boulevard with a rooftop promenade offering panoramic views. These features are designed to promote physical well-being and social interaction in an accessible, low-tourism environment.

3.3 T-Factor Public–Private Partnership to Map Urban Biodiversity

T-Factor is a Horizon 2020 initiative that redefines temporary urbanism as a tool for early-stage regeneration. It fosters innovation, collaboration, and policy-making to create inclusive, sustainable, and community-driven urban transformations.

T-Factor operates across six early-stage regeneration sites in Europe, leveraging temporary uses to activate public spaces and shape future urban development. By integrating cities, universities, businesses, and grassroots organizations, it promotes knowledge sharing and best practices in urban transformation. With 25 international partners, T-Factor explores the role of Thematic Labs (T-Labs), covering areas such as Arts and Creativity, Urban Design, Circular Economy, and Climate Change. These multidisciplinary approaches contribute to policy innovation and help frame temporary urbanism as a structured process, bridging short-term interventions with long-term strategic urban development. Within this research framework, T-Factor investigated the role of Nature in the design of meanwhile landscapes during urban regeneration, through biodiversity mapping actions and processes developed in the Bioscopium initiative.

Bioscopium was conceived as one of the initiatives developed by T-Factor for the pilot project of Herbula Wild Garden at MIND (Milano Innovation District), carried out by the designers of LAND, researchers of Polifactory (Department of Design, Politecnico di Milano) in collaboration with agronomists and wildlife experts of StudioTerraViva [15]. In Herbula Wild Garden, Bioscopium was tested and developed to introduce **an innovative approach to exploring urban biodiversity within areas undergoing regeneration**.

Bioscopium integrates four main components: GIS-based ecological network mapping, landscape indicator monitoring, citizen science engagement, and modular sensor stations inspired by AMMOD technology. Together, these tools enable a dynamic understanding of how regeneration impacts biodiversity, while identifying ways to enhance it strategically over time.

Its main purpose is to offer decision-makers a practical tool to assess the ecological impact of transformation processes, while also supporting specialists in mapping and monitoring efforts. A key element of the system is the active involvement of citizens and local stakeholders in biodiversity-related initiatives and the continuous dialogue with private and public entities to establish a legacy based on the promotion and enhancement of local biodiversity among real estate developers, NGOs, local municipalities, and citizens.

A key result of the project is a set of practical guidelines to make biodiversity mapping a standard part of regeneration efforts. These guidelines support a wide range of stakeholders (from developers and public authorities to ecologists, designers, and citizens) by offering tools, methods, and actions for integrating biodiversity into planning, policy, and community stewardship.

Fig. 3 Herbula Wild Garden, MIND (Milano Innovation District). Image from Polifactory, *T-Factor*. *Source* polifactory.polimi.it

They emphasize awareness-raising, ecological monitoring, use of open-source tools, and collaborative ecosystems to promote biodiversity as a core value of equitable, resilient urban development. Tested at MIND in Milan, Bioscopium lays the groundwork for replicable, community-driven models that embed nature at the center of urban transformation (Fig. 3).

4 Conclusions and Further Outcomes

To advance a Nature-Positive vision in urban regeneration, several actionable strategies emerge from the projects in Essen, Riyadh, and Milan.

First, there is a need to reframe nature not as a constraint, but as vital infrastructure—essential for public health, climate adaptation, and social well-being. Urban regeneration processes, particularly on brownfield or underused sites, offer strategic opportunities to reintroduce biodiversity and strengthen ecological networks, reversing degradation and restoring continuity across landscapes' scales. In this context, embedding biodiversity mapping into early planning stages becomes crucial. Tools such as GIS-based analyses, sensor technologies, and citizen science not only provide data to inform design decisions but also support adaptive strategies that evolve over time. By standardizing these methodologies, cities can ensure that ecological knowledge is consistently translated into design practice.

Equally important is the fostering of public–private partnerships. Aligning the interests of developers, local authorities, and communities can help mobilize collective investment in green infrastructure, where private sector productivity supports public goods such as environmental restoration, resilience, and access to quality public space. Design solutions must also address climate resilience by incorporating sustainable systems—such as rainwater harvesting, native vegetation, and reforestation—to manage extreme temperatures and water scarcity, particularly in vulnerable urban environments. Community engagement is a cornerstone of this process. Involving citizens through participatory tools and citizen science initiatives helps build a shared sense of responsibility for urban nature, transforming public space into an active site of ecological learning and stewardship.

Finally, it is essential to translate broad policy ambitions into tangible design outcomes. Frameworks like the *Nature-Factory Manifesto* and tools such as Natural Capital Accounting (United Nations 2019) help bridge this gap by offering operational guidelines that align global sustainability goals with local, context-specific implementation. By documenting and sharing these practices, cities and practitioners can create scalable, transferable models that adapt Nature-Positive principles to diverse geographic and cultural realities ensuring that ecological regeneration becomes the norm, not the exception, in the urban landscape.

References

1. United Nations Convention on Biological Diversity (2022) Kunming-Montreal global biodiversity framework. Retrieved from https://www.cbd.int/gbf/
2. Taskforce on Nature-related Financial Disclosures (2023) TNFD beta framework v0.4. Retrieved from https://tnfd.global
3. Science Based Targets Network (2023) Guidance for setting science-based targets for nature. Retrieved from https://sciencebasedtargetsnetwork.org
4. UK Department for Environment, Food & Rural Affairs (2023) Statutory biodiversity metric guidelines. Retrieved from https://www.gov.uk/government/publications/
5. Italian Ministry for the Environment and Energy Security (2022) Strategia Nazionale per la Biodiversità 2030. Retrieved from https://www.mase.gov.it
6. McDonald R et al (2019) Nature in the urban century: a global assessment of where and how to conserve nature for biodiversity and human wellbeing. The Nature Conservancy
7. OECD (2020) A territorial approach to the sustainable development goals: synthesis report. Retrieved from https://www.oecd.org
8. City of Durban (2022) eThekwini strategic hub (StratHub) dashboard and biodiversity action framework. Retrieved from https://www.durban.gov.za
9. CDP (2023) Protecting people and the planet: how cities can lead the way to a nature-positive economy. Retrieved from https://www.cdp.net
10. Intergovernmental Science-Policy Platform on Biodiversity and Ecosystem Services (IPBES) (2019) Global assessment report on biodiversity and ecosystem services. Retrieved from https://ipbes.net/global-assessment

11. Nature Positive Cities (2025) Nature positive cities initiative. Retrieved from https://www.Naturepositivecities.org/home
12. LAND (2025) The nature-factory manifesto. Retrieved from https://www.landsrl.com/wp-content/uploads/2025/01/Nature-factory-manifesto-def.pdf
13. LAND (2010) Neue Wege zum Wasser, Essen. Retrieved from https://www.landsrl.com/en/work/neue-wege-zum-wasser-essen/
14. Saudi vision 2030 (2025) Green Riyadh project. Retrieved from https://www.vision2030.gov.sa/en/explore/projects/green-riyadh
15. Polifactory (Politecnico di Milano), LAND Italia (2024) Bioscopium: guidelines for mapping urban biodiversity in urban regeneration areas. T-Factor. Retrieved from https://www.t-factor.eu/bioscopium-handbook/

Open Access This chapter is licensed under the terms of the Creative Commons Attribution-NonCommercial-NoDerivatives 4.0 International License (http://creativecommons.org/licenses/by-nc-nd/4.0/), which permits any noncommercial use, sharing, distribution and reproduction in any medium or format, as long as you give appropriate credit to the original author(s) and the source, provide a link to the Creative Commons license and indicate if you modified the licensed material. You do not have permission under this license to share adapted material derived from this chapter or parts of it.

The images or other third party material in this chapter are included in the chapter's Creative Commons license, unless indicated otherwise in a credit line to the material. If material is not included in the chapter's Creative Commons license and your intended use is not permitted by statutory regulation or exceeds the permitted use, you will need to obtain permission directly from the copyright holder.

Role of Spatial Planning in Addressing Climate Challenges: A Study Concerning the Functional Urban Area of Cagliari

Sabrina Lai and Corrado Zoppi

Keywords Ecosystem services · Urban planning · Carbon sequestration and storage · GIS · Inferential models

1 Introduction

The capacity for carbon sequestration (CSC) enhances the absorption of carbon in natural environments, whereas disposal methods inhibit carbon dioxide (CO_2) from returning to the active carbon cycle. These techniques are typically organized according to the Earth system component they aim to influence: land-based CSC, marine CSC, and geological storage [1]. Land-based CSC utilizes photosynthesis to transform atmospheric CO_2 into organic matter and soil-based carbon stocks, offering co-benefits like enriched soil fertility, higher agricultural productivity, and cleaner water resources [2].

Given that CO_2 is the principal greenhouse gas accelerating climate change, a range of engineering and ecological strategies has been introduced to curb its emissions and maintain stable concentrations in the atmosphere. Among these, natural CSC boosts the capacity of ecosystems such as woodlands and soils to capture and retain carbon [3]. Strengthening these natural carbon sinks should thus be a central consideration in planning land use.

S. Lai (✉) · C. Zoppi
Dipartimento di Ingegneria Civile, Ambientale e Architettura (DICAAR), University of Cagliari, Cagliari, Italy
e-mail: sabrinalai@unica.it

C. Zoppi
e-mail: zoppi@unica.it

This research defines and applies a methodological framework aimed at achieving climate neutrality through spatial planning mechanisms. CSC is employed as a core analytical concept to meet this aim, grounded in an array of ecosystem services (ESs). The study unfolds in several stages. Initially, the spatial dimensions of CSC are assessed using CSC density mapping through the "Carbon Storage and Sequestration" module from the InVEST toolset [4], which calculates carbon reserves across land units based on land cover data [5].

In the next phase, the relationship between CSC and various ESs is modeled and spatially evaluated, focusing on services such as sustaining biodiversity by ensuring viable habitats for flora and fauna beneficial to humans; moderating climate by lowering land surface temperature (LSTE); controlling water runoff; and providing areas for recreational activities in natural settings.

Lastly, spatial correlations between CSC metrics and ES categories are examined within the Functional Urban Area (FUA) of Cagliari, situated in Sardinia, an island in Southern Italy. This analysis aims to determine how ES provision can be strategically optimized to enhance CSC performance while refining the spatial arrangement of ESs. The findings contribute to place-sensitive policy guidance for improving environmental quality through better ES delivery in FUAs.

2 Methodologies and Materials

2.1 The FUA of Cagliari

The FUA of Cagliari (Fig. 1), located in Sardinia, Italy, spans approximately 2000 km^2, includes 32 municipalities, and has a population of 475,170 as of 2023. To calculate the model variables detailed in Sect. 2.3, a vector-based grid of ~200,000 cells, each measuring 100 × 100 m, was developed to cover the FUA.

2.2 CSC Definition and Spatial Taxonomies of ESs

CSC is in this study was analyzed alongside four additional ESs relevant to the FUA: nature-based recreation, pluvial runoff regulation, habitat quality, and LST regulation. Table 1 lists the spatial datasets and sources used to evaluate the density of carbon stored in carbon sinks (CSCD) and the other ESs.

For CSCD, InVEST's [4] "Carbon Storage and Sequestration" module generates a raster map representing carbon density per pixel, calculated as the sum of carbon stored in four pools: aboveground biomass, belowground biomass, dead organic matter, and soil organic carbon. This output is derived from a land cover dataset and a corresponding look-up table assigning carbon values to each land cover type. Due to the absence of data on belowground biomass, the model was run using only the

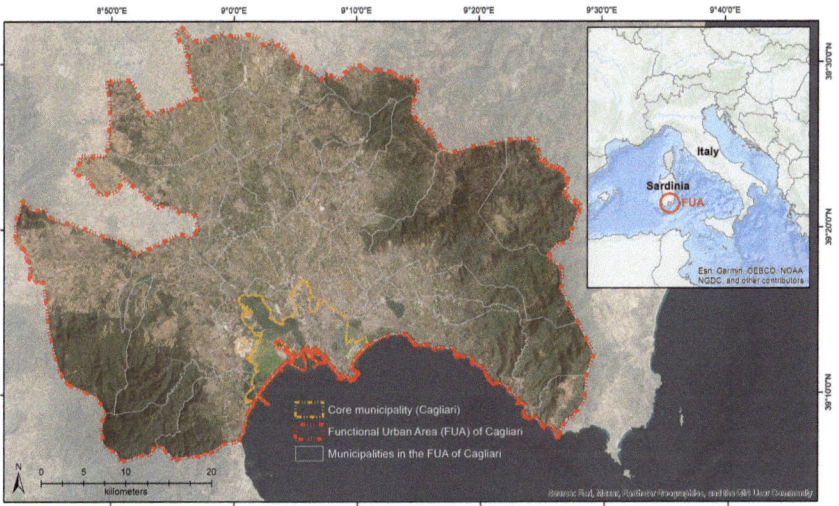

Fig. 1 Study area: the functional urban area of Cagliari and its core municipality

remaining three pools. Carbon values were sourced from the 2005 Italian forest and carbon sink inventory and supplemented by field data collected through a regional pilot initiative led by two Sardinian agencies specializing in rural development and agricultural research.

RCTR was evaluated using InVEST's [4] "Urban Flood Risk Mitigation" model, which requires the following inputs: i., a land cover dataset; ii., a soil hydrologic group map based on USDA-NRCS's [16] classification; iii., a biophysical table linking curve numbers to each land cover–soil group combination; iv., rainfall depth; and v., watershed boundaries for spatial aggregation. The model outputs a raster map of runoff retention, where water retained per pixel is determined by rainfall and runoff, both influenced by land use and soil permeability. All input data were sourced from regional repositories.

HAQU was assessed using InVEST's [4] "Habitat Quality" model, which evaluates ecosystem integrity based on degradation caused by the presence and intensity of biodiversity threats and considering institutional protection, which may limit exposure. The model uses a raster land cover map to represent habitat distribution, supplemented by a table assigning habitat suitability scores to each land cover type (ranging from 0 to 1). In this study, the CORINE 2018 land cover dataset was employed. Threat data were obtained from the Natura 2000 standard data forms and the regional geoportal. Two expert-informed tables were developed: one defining the severity and spatial influence of each threat, and another assigning suitability values and sensitivity scores to each land cover type.

LSTE was derived from raster maps of LST, used as a proxy for ecosystems' temperature-regulating capacity. Landsat Collection 2, Level 2 imagery (May–October 2023) was sourced from the USGS, excluding scenes with over 10% cloud

Table 1 Data used to map the variables used in this study

Variable	Data requirements	Source
CSCD	Land cover map	Sardegna Geoportale [6]
	Reference table linking land cover categories to carbon pools data	Sardegna Geoportale [6, Inventario Nazionale delle Foreste e dei serbatoi forestali di Carbonio [National Inventory of Forests and forest Carbon pools] 7]
RECA	Land cover map	Urban Atlas [8]
	Vector map of census tracts Table linking census tract identifiers to population data	Dati per sezione di censimento [Data by Census Tract] [9]
RCTR	Land cover map	Sardegna Geoportale [6]
	Map outlining watershed regions	Sardegna Geoportale [6]
	Map of soil permeability characteristics	Sardegna Geoportale [6]
	Table mapping land cover types to curve number values	Carta del curve number regionale [Regional Curve Number Map] [10]
	Data on rainfall	Idrologia e Idrometria [Hydrology and Hydrometry] [11]
HAQU	Land cover map (CORINE 2018)	CORINE Land Cover [12]
	List and geographic datasets of environmental threats	Sardegna Geoportale [6, Natura 2000 Viewer 13]
	Table with information on threats	Lai and Leone [14]
	Table indicating sensitivity to various threats	
	Accessibility to threats	Sardegna Geoportale [6]
LSTE	Satellite imagery Landsat C2L2	EarthExplorer [15]

cover. The image with the highest mean temperature, dated July 30, 2023, was selected for analysis.

RECA captures the flow of nature-based recreational opportunities. Supply was measured as the availability of green and blue spaces, while demand was estimated as the number of residents within 500 m of these areas. The indicator was computed by multiplying the share of recreational land within each grid cell by the local population within a 500-m radius. Recreation-suitable areas were identified using the 2018 Urban Atlas, and population data were obtained from the 2021 national census, including tract boundaries and resident counts. For detailed methodology, see [17].

After generating the raster maps for CSCD and the other four ESs, mean values for each 100-m square cell were computed using zonal statistics.

To control for the effect of soil presence and consistency, a variable named LCAP was introduced in the regression model; it equals 1 for arable soils (classes I–IV) and 0 for non-arable soils (classes V–VIII), following a classification by USDA [18] and using as input a regional soil map of Sardinia, reclassified and aggregated to the grid level via zonal statistics. An additional control variable, CSCL, was included to account for spatial autocorrelation in CSCD. Defined as the spatial lag of CSCD, CSCL was computed using Moran's I in GeoDa [19], applying a first-order queen contiguity weight matrix to the 100-m vector grid.

2.3 Linear Regression

The correlations between the spatial taxonomy of CSC and the supply of the selected ESs are estimated through a linear regression, which develops as follows:

$$CSCD = \beta_0 + \beta_1 RECA + \beta_2 RCTR + \beta_3 HAQU + \beta_4 LSTE + \beta_5 LCAP + \beta_6 CSCL \quad (1)$$

The dependent and covariates are the following, which relate to a cell of 100×100 m:

- CSCD is CSC density ($Mg/(100 \text{ m}^2)$);
- RECA is the percentage of recreational areas for outdoor activities times the residents located within a 500-m radius around a cell (% times residents);
- RCTR is the amount of precipitation that can be absorbed or held, and thus does not contribute to surface runoff (m^3);
- HAQU is habitat quality (this variable takes rational values between 0 and 1; see Sect. 2.2);
- LSTE is taken as a proxy for heat mitigation size (°C);
- LCAP is a Boolean variable which has value 1 for arable soils and 0 otherwise; for the definition see Sect. 2.2);
- CSCL is a control variable related to spatial autocorrelation of CSCD.

Linear regression is used to assess the relationship between CSC spatial patterns and selected ESs supply. Predictive models based on regression are commonly applied in situations where prior knowledge about the interconnections between variables describing intricate systems is lacking [20, 21]. Coefficients quantify links between CSC distribution, ESs provision, and control factors. LCAP, reflecting soil arability, accounts for CSC reductions in arable soil due to organic carbon loss [22]. Spatial autocorrelation in CSC is addressed using the spatial lag variable CSCL, following Anselin's approach [19, 23]. Model significance is tested via p-values for all variables.

3 Results

3.1 CSC Definition and Spatial Taxonomies of ESs

Figure 2 presents the spatial distribution of CSCD, RECA, and RCTR. As for CSCD (panel A), the highest values, falling within the top two deciles, are primarily located along the FUA edges, where extensive forest cover and protected natural areas dominate. These include major Natura 2000 sites and regional parks. In contrast, the urban core of Cagliari and adjacent wetlands consistently show the lowest CSCD values, while agricultural zones register intermediate levels. Concerning RECA (panel B), the highest decile values are concentrated in the central FUA, particularly in urbanized areas where green and blue spaces are accessible within 500 m of residential zones. To the west, medium-to-high values appear in low-density areas near the "Sette Fratelli" forest. Conversely, a prominent low-value cluster in the east is due to the absence of nearby settlements. The lowest RCTR values (panel C) correspond to impervious surfaces primarily concentrated in and around the FUA's urban core and other settlements. Conversely, the highest values (ninth and tenth deciles) are located along the FUA's western and eastern boundaries. Moderate-to-high retention values (sixth and seventh deciles) are mainly found in permeable agricultural zones.

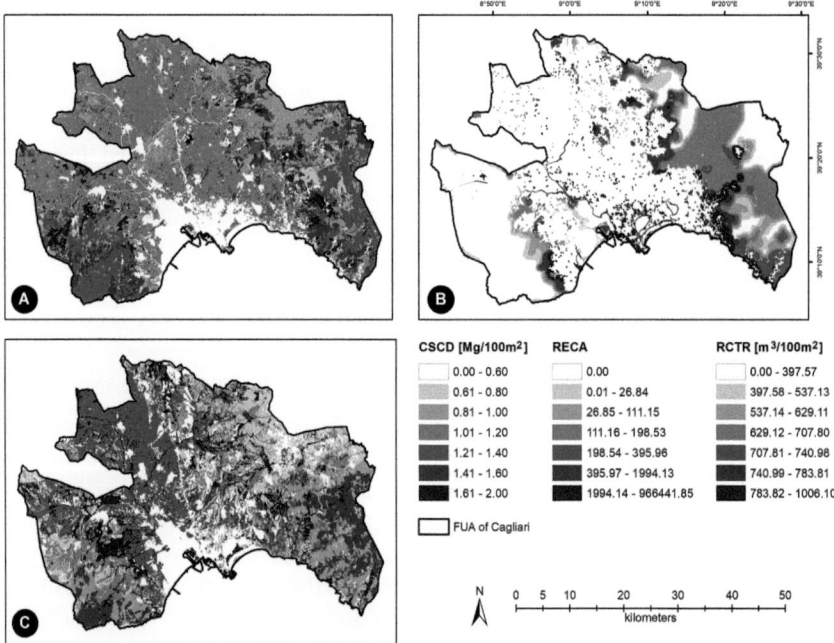

Fig. 2 Spatial layout of CSCD (panel A), RECA (panel B), and RCTR (panel C) in the FUA

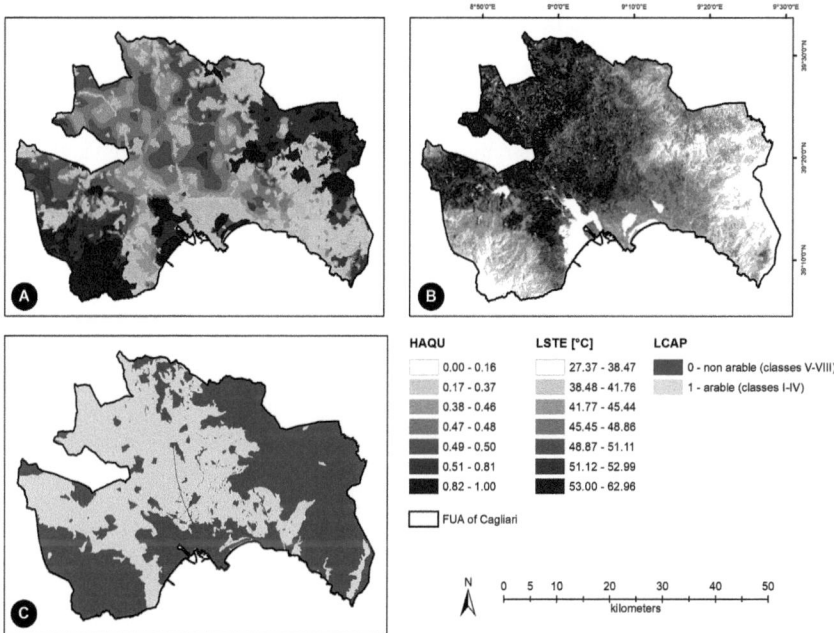

Fig. 3 Spatial layout of HAQU (panel A), LSTE (panel B), and LCAP (panel C) in the FUA

Figure 3 shows the distribution of HAQU, LSTE, and LCAP. As for HAQU (panel A), the highest scores, in the top two deciles, appear not only along the forested and protected western and eastern margins, but also near the urban core and its surroundings. This pattern reflects the significant biodiversity and legal protections of two large Natura 2000 wetlands. Areas in the lowest decile generally correspond to artificial land covers, while rural zones display a gradient from low to medium–high values. Concerning LSTE (panel B), the spatial variation of LST recorded on the hottest summer day of 2023 is shown. LSTE indicates ecosystems' ability to moderate local temperatures, where higher LST values correspond to reduced regulating capacity. The coolest temperatures are found in inland water bodies (first decile), followed by elevated terrain (second and third deciles). The highest temperatures occur across the agricultural Campidano Plain. The Cagliari core and adjacent areas show moderate values, likely influenced by the sea and the two wetlands.

3.2 Linear Regression

Model (1) estimates, shown in Table 2, reveal how CSCD responds to changes in four ES supply and control variables. Each coefficient indicates the marginal effect on CSCD per unit increase in the respective variable, with significance confirmed via p-values.

A positive coefficient means that higher ES supply increases CSCD, and vice versa. All estimates are statistically significant, making them reliable indicators of the relationships analyzed. The strength of each effect is expressed, in sign and size, through elasticity, calculated as the ratio of percentage changes in CSCD to those in each variable. Table 2's final column presents these elasticity values, related to the mean values of the covariates. Since all values are below 100%, CSCD is generally inelastic to changes in ES supply, though variation exists, in size and sign, across ESs.

According to model (1), despite significance, the four ESs affect CSCD differently. A 5% increase in LSTE (about a 2.5 °C rise) leads to a ~1% drop in CSCD, indicating a weak negative effect due to LSTE's typically small temporal variation.

HAQU shows a smaller positive elasticity than LSTE's negative one, suggesting improvements in habitat quality have limited effect. For instance, a 10% HAQU increase, related, for example, to a relevant decline in the size of threats to habitat quality in the FUA of Cagliari, yields under 1% CSCD growth.

RECA negatively affects CSCD but has a very low elasticity (~0.4%). However, as for the definition of RECA delivered in Sects. 2.2 and 2.3, a 10% drop in buffer-zone population could result in a ~10% CSCD increase, implying a substantial negative elasticity (~100%). Still, since population shifts are minimal over time, the short- to medium-term effect is weak.

RCTR has the strongest positive influence on CSCD, with an elasticity close to 50%, despite CSCD's overall inelasticity.

LCAP negatively impacts CSCD; a 10% rise in arable land (~5% of the total) reduces CSCD by ~0.2%.

Table 2 Regression results

Explanatory variable	Coefficient	t-statistic	p-value	Mean of the explanatory variable	Elasticity at the mean values of CSCD and expl. var's, related to a 10% increase in expl. var's [$(\Delta y/y)/(\Delta x/x)$, %]
RECA	−2.0E−6	−66.7691	0.000	2074.3821	−0.3897
RCTR	0.0008	220.3108	0.000	628.7480	48.2439
HAQU	0.2265	86.5215	0.000	0.4571	9.7642
LSTE	−0.0047	−31.5592	0.000	46.4666	−20.7214
LCAP	−0.0527	−29.57889	0.000	0.4781	−2.3771
CSCL	0.3136	35.8959	0.000	0.2568	7.5937

Dependent variable: CSCD: Mean: 1.0604 Mg/(100 m^2); Standard deviation: 0.3270 Mg/(100 m^2); Adjusted R-squared: 0.2986

Finally, CSCL confirms significant spatial autocorrelation, reinforcing spatial dependency in CSC distribution.

4 Discussion

The relationship between CSC and LSTE, as shown through the spatial analysis of the Cagliari FUA, is consistent with numerous studies. Momo and Devi [24] analyzed LST and CSC trends over 2011–2021 in Imphal's West District, India, using various satellite-based methods. Their findings showed consistent declines in CSC and rising LST. Likewise, Wang et al. [25] studied subtropical Shenzhen, China, where urban heat islands led to lower CSC levels, especially in dense urban cores compared to peripheral areas.

Many studies also highlight the positive link between CSC and habitat quality across spatial scales, local to continental. CSC is considered a key structural aspect of habitat quality, actively enhancing environmental health. Bayley et al. [26], for example, found improved CS tied to forest restoration on the Falkland Islands. In urban Xiamen, Hua et al. [27] introduced a model assessing trade-offs from urban growth, showing how services like CSC, water regulation, habitat quality, and soil stability change in coordinated ways.

A negative correlation between CSC and population density is also widely observed. Kinnunen et al. [28] note that dense urban areas contribute little to CSC, with declining capacity linked to building density. Gao and O'Neill [29] emphasize that population growth and urban sprawl increase CO_2 emissions and reduce high-value natural carbon sinks. Riparian zones and wetlands still serve as strong carbon sinks. Pan et al. [30] point out that forest root systems enhance CSC and water retention, aiding flood control. Kumar et al. [31] observe this effect in western India, where afforestation improves both CSC and hydrological resilience. In cities, green roofs provide similar benefits, improving both stormwater control and CSC through vegetative root systems [32].

5 Conclusion

This study has presented a novel method combining geospatial analysis and modeling to examine links between CS and other ESs in the Cagliari FUA, focusing on runoff retention, temperature regulation, habitat quality, and nature-based recreation.

Quantifying these relationships supports evidence-based urban planning for permeable soils and green spaces. Key urban strategies include soil desealing, afforestation, and green space expansion, primarily in public areas, where private property regulations should encourage maintaining unsealed surfaces and appropriate vegetation. In agricultural areas, however, management practices are often more critical than planning policies for enhancing carbon sequestration and related

ESs. This approach is notable for its simplicity and adaptability, using standard inferential models on publicly available biophysical spatial data. Data limitations, such as missing belowground carbon and coarse soil maps, pose challenges; therefore, future work could consider using more detailed datasets. Moreover, the range of ESs studied could be broadened to better inform planning and policy decisions.

Acknowledgements This study was carried out: i. within the RETURN Extended Partnership funded by the European Union—NextGenerationEU (National Recovery and Resilience Plan—NRRP, M4, C2, In-VESTment 1.3—D.D. 1243 2/8/2022, PE0000005); ii. within the NRRP, M4, C2, In-VESTment 1.1, Call for tender no. 1409 published on 14.9.2022 by the Italian Ministry of University and Research (MUR), funded by the European Union—NextGenerationEU—Project Title "Definition of a guidelines handbook to implement climate neutrality by improving ecosystem service effectiveness in rural and urban areas"—CUP F53D23010760001—Grant Assignment Decree no. 1378 adopted on September 1, 2023, by MUR; iii. within the research grant CUP F73C23001680007 for the project "Geodesign for climate change mitigation and adaptation in the Mediterranean region," funded in 2022 by Fondazione di Sardegna.

Authors' Contributions Sabrina (S.L.) and Corrado Zoppi (C.Z.) collaboratively designed this study. Individual contributions are as follows: C.Z. wrote Sections 1, 2.3, 3.2 and 4; S.L. wrote Sectios 2.1, 2.2, 3.1 and 5.

Disclosure of Interests The authors declare no competing interests.

References

1. Dilling L, Doney SC, Edmonds J, Gurney KR, Harriss R, Schimel D, Stephens B, Stokes G (2003) The role of carbon cycle observations and knowledge in carbon management. Annu Rev Environ Res 28(1):521–558. https://doi.org/10.1146/annurev.energy.28.011503.163443
2. Lal R (2008) Carbon sequestration. Philos Trans R Soc B 363(1492):815–830. https://doi.org/10.1098/rstb.2007.2185
3. Ghommem M, Hajj MR, Puri IK (2012) Influence of natural and anthropogenic carbon dioxide sequestration on global warming. Ecol. Modell. 235–236:1–7. https://doi.org/10.1016/j.ecolmodel.2012.04.005
4. What is InVEST. https://naturalcapitalproject.stanford.edu/software/invest. Accessed 20 May 2025
5. Liquete C, Kleeschulte S, Dige G, Maes J, Grizzetti B, Olah B, Zulian G (2015) Mapping green infrastructure based on ecosystem services and ecological networks: a pan-European case study. Environ Sci Policy 54:268–280. https://doi.org/10.1016/j.envsci.2015.07.009
6. SardegnaGeoportale. https://www.sardegnageoportale.it/. Accessed 20 May 2025
7. Inventario Nazionale delle Foreste e dei serbatoi forestali di Carbonio [National Inventory of Forests and forest Carbon pools]. https://www.inventarioforestale.org/en/. Accessed 20 May 2025
8. Urban Atlas. https://land.copernicus.eu/en/products/urban-atlas. Accessed 20 May 2025
9. Dati per sezione di censimento [Data by Census Tract]. https://www.istat.it/it/archivio/285267. Accessed 20 May 2025
10. Carta del curve number regionale [Regional Curve Number Map]. https://www.sardegnageoportale.it/documenti/40_615_20190329081206.pdf. Accessed 20 May 2025
11. Idrologia e Idrometria [Hydrology and Hydrometry]. https://www.sardegnaambiente.it/index.php?xsl=611&s=21&v=9&c=93749&na=1&n=10. Accessed 20 May 2025

12. CORINE Land Cover. https://land.copernicus.eu/en/products/corine-land-cover/clc2018. Accessed 20 May 2025
13. Natura 2000 Viewer. https://natura2000.eea.europa.eu. Accessed 20 May 2025
14. Lai S, Leone F (2017) Bridging biodiversity conservation objectives with landscape planning through green infrastructures: a case study from Sardinia, Italy. In: Gervasi O et al (eds) 17th international conference on computational science and its applications (ICCSA 2017), Lecture notes in computer sciences series (LNCS), vol 10409, pp 456–472. Springer, Cham. https://doi.org/10.1007/978-3-319-62407-5_32
15. EarthExplorer. https://earthexplorer.usgs.gov. Accessed 20 May 2025
16. USDA-NRCS (2009) National engineering handbook. Part 630 Hydrology. Chapter 7 Hydrologic soil groups. https://www.hydrocad.net/neh/630ch7.pdf. Accessed 20 May 2025
17. Isola F, Lai S, Leone F, Zoppi C (2024) Urban green infrastructure and ecosystem service supply: a study concerning the functional urban area of Cagliari, Italy. Sustainability 16(19):8628. https://doi.org/10.3390/su16198628
18. Klingebiel AA, Montgomery PH (1961) Land capability classification. https://www.gov info.gov/content/pkg/GOVPUB-A-PURL-gpo20777/pdf/GOVPUB-A-PURL-gpo20777.pdf. Accessed 20 May 2025
19. Anselin L, Syabri I, Kho Y (2006) GeoDa: an introduction to spatial data analysis. Geogr Anal 38(1):5–22. https://doi.org/10.1111/j.0016-7363.2005.00671.x
20. Sklenicka P, Molnarova K, Pixova KC, Salek ME (2013) Factors affecting farmlands in the Czech Republic. Land Use Policy 30(1):130–136. https://doi.org/10.1016/j.landusepol.2012.03.005
21. Stewart PA, Libby LW (1998) Determinants of farmland value: the case of DeKalb County, Illinois. Rev Agric Econ 20(1):80–95. https://www.jstor.org/stable/1349535. Accessed 20 May 2025
22. Anuo CO, Sleem M, Fossum B, Li L, Cooper JA, Malakar A, Maharjan B, Kaiser M (2024) Land use selectively impacts soil carbon storage in particulate, water-extractable, and mineral-associated forms across pedogenetic horizons. Geoderma 449:116992. https://doi.org/10.1016/j.geoderma.2024.116992
23. Anselin L (2003) Spatial econometrics. In: Baltagi BH (ed) A companion to theoretical econometrics. Blackwell Publishing, Oxford, pp 310–330
24. Momo M, Devi TT (2022) Assessment of land surface temperature and carbon sequestration using remotely sensed satellite data in the Imphal-West District, Manipur, India. J Earth Syst Sci 131:229. https://doi.org/10.1007/s12040-022-01944-8
25. Wang J, Xiang Z, Wang W, Chang W, Wang Y (2021) Impacts of strengthened warming by urban heat island on carbon sequestration of urban ecosystems in a subtropical city of China. Urban Ecosyst 24:1165–1177. https://doi.org/10.1007/s11252-021-01104-8
26. Bayley DTI, Brickle P, Brewin PE, Golding N, Pelembe T (2021) Valuation of kelp forest ecosystem service in the Falkland islands: a case study integrating blue carbon sequestration potential. One Ecosyst 6:e62811. https://doi.org/10.3897/oneeco.6.e62811
27. Hua Y, Yan D, Liu X (2024) Assessing synergies and trade-offs between ecosystem services in highly urbanized area under different scenarios of future land use change. Environ Sustain Indic 22:100350. https://doi.org/10.1016/j.indic.2024.100350
28. Kinnunen A, Talvitie I, Ottelin J, Heinonen J, Junnila S (2022) Carbon sequestration and storage potential of urban residential environment—a review. Sustain Cities Soc 84:104027. https://doi.org/10.1016/j.scs.2022.104027
29. Gao J, O'Neill B (2020) Mapping global urban land for the 21st century with data-driven simulations and shared socioeconomic pathways. Nat Commun 11:2302. https://doi.org/10.1038/s41467-020-15788-7
30. Pan C, Shrestha A, Innes JL, Zhou G, Li N, Li J, He Y, Sheng C, Niles J-O, Wang G (2022) Key challenges and approaches to addressing barriers in forest carbon offset projects. J For Res 33:1109–1122. https://doi.org/10.1007/s11676-022-01488-z

31. Kumar R, Bhatnagar PR, Kakade V, Dobhal S (2020) Tree plantation and soil water conservation enhances climate resilience and carbon sequestration of agro ecosystem in semi-arid degraded ravine lands. Agr Forest Meteorol 282–283:107857. https://doi.org/10.1016/j.agrformet.2019.107857
32. Mihalakakou G, Souliotis M, Papadaki M, Menounou P, Dimopoulos P, Kolokotsa D, Paravantis JA, Tsangrassoulis A, Panaras G, Giannakopoulos E, Papaefthimiou S (2023) Green roofs as a nature-based solution for improving urban sustainability: progress and perspectives. Renew Sust Energ Rev 180:113306. https://doi.org/10.1016/j.rser.2023.113306

Open Access This chapter is licensed under the terms of the Creative Commons Attribution-NonCommercial-NoDerivatives 4.0 International License (http://creativecommons.org/licenses/by-nc-nd/4.0/), which permits any noncommercial use, sharing, distribution and reproduction in any medium or format, as long as you give appropriate credit to the original author(s) and the source, provide a link to the Creative Commons license and indicate if you modified the licensed material. You do not have permission under this license to share adapted material derived from this chapter or parts of it.

The images or other third party material in this chapter are included in the chapter's Creative Commons license, unless indicated otherwise in a credit line to the material. If material is not included in the chapter's Creative Commons license and your intended use is not permitted by statutory regulation or exceeds the permitted use, you will need to obtain permission directly from the copyright holder.

Planning for Biodiversity: Strategies and Actions for Enhancing Nature in the Urban Plan of Varese (Italy)

Andrea Arcidiacono⊙, Laura Pogliani⊙, Silvia Ronchi⊙, Stefano Salata⊙, Andrea Benedini⊙, Federico Ghirardelli, and Beatrice Mosso⊙

Keywords Ecosystem services · Urban planning · Urban forestry · Nature restoration · Green and blue infrastructures

1 Introduction

Recent urbanisation and climate change are pushing many species to the brink of extinction [1]. Urban areas are expected to triple by 2030 [2, 3] with the majority of urbanisation taking place in biodiversity hotspots [4], threatening global biodiversity and reducing the Ecosystem Services (ES) capacity [5]. The link between biodiversity and cities has historical roots, based on the early awareness of nature's positive effect on human health and well-being [6]. The rising focus on urban biodiversity protection has resulted in the development of many international and European policy frameworks. Starting from the Convention on Biological Diversity (1972), which highlights the role of cities in biodiversity conservation, at the European level, the EU Green Deal places biodiversity at the core of sustainable development strategies [7, 8]. Additional initiatives, such as the Green and Blue Infrastructures (GBI) strategy, promote the use of Nature-based Solutions (NbS) for their potential to address biodiversity and climate crises [9–11].

Furthermore, the EU Biodiversity Strategy for 2030 is a long-term plan to protect nature and reverse ecosystem degradation. It recognises the crucial role of cities

---

A. Arcidiacono · L. Pogliani · S. Ronchi (✉) · S. Salata · A. Benedini · F. Ghirardelli · B. Mosso
Department of Architecture and Urban Studies (DAStU), Politecnico di Milano, Milano, Italy
e-mail: silvia.ronchi@polimi.it

A. Arcidiacono · S. Ronchi
National Biodiversity Future Center (NBFC), Palermo, Italy

in achieving these targets, promotes a strong increase in green areas, and sets the ambitious target of planting at least 3 billion trees by 2030 [12].

The recently proposed Nature Restoration Law strengthens this approach by establishing legally binding restoration targets [13]. It requires Member States to prioritise the recovery of urban ecosystems, including restoring damaged habitats and expanding biological corridors within cities [14]. According to the proposed law, cities with over 20,000 inhabitants must adopt 'Urban Greening Plans' to increase green spaces, enhance ecological connectivity, and encourage NbS within the urban fabric.

These biodiversity targets align with the EU's goal of No Net Land Take (NNLT) by 2050, as outlined in the Soil Strategy for 2030 [15], aiming to balance new urban development with land restoration by preventing soil loss for nature or agriculture while prioritising urban regeneration.

The achievement of European objectives requires the adoption of a combined strategy that aims to reduce future land take (avoiding the degradation of further habitats) and promote de-sealing initiatives (which are thus essential strategies to reclaim permeable surfaces and create habitats that support urban biodiversity) [16].

Traditionally, urban green strategies are often carried out using sectoral plans and policies without real integration into comprehensive land-use planning. Moreover, these strategies are focused mainly on increasing the quantity of green spaces without using an integrated design approach that also considers the quality of these areas in terms of ES performance [17].

Biodiversity interventions, when aligned with GBI and NbS, significantly enhance the provision of ES, moving beyond isolated greening projects to foster interconnected networks of multiple areas capable of sustaining biodiversity across urban landscapes [18]. This requires overcoming the prevailing compartmentalisation of planning practices and embracing integrative, multi-scalar approaches that align NbS, GBI, and biodiversity objectives within comprehensive urban strategies.

According to this framework, limiting land take also entails implementing a downzoning strategy, which actively emerges as a strategic planning tool to prevent future land take. This strategy may include lowering building rights or development potential in previously developable areas [19]. By redesignating areas to lower-intensity uses or ecological functions, downzoning can directly support the spatial design of GBI. So, incorporating downzoning within the GBI planning process enables linking regulatory land-use measures with ecological objectives, making GBI the structural backbone of urban development and, at the same time, the operational tool for achieving NNLT and biodiversity goals [20].

Additionally, GBI should integrate desealing strategies to improve biodiversity connectivity and restore ecosystem functionality, converting impermeable urban surfaces into multipurpose green areas that provide essential ES [16].

The case study of Varese illustrates how the design of an ES-based GBI can effectively orient planning decisions toward biodiversity enhancement, as required by current EU policies. In particular, a downzoning strategy was adopted to limit future land-take, increase biodiversity based on ES performance, and promote the restoration of low-performing urban fabric.

2 Varese Garden City

Varese is a medium-sized city of approximately 78,000 inhabitants in northern Italy's Lombardy region, roughly 55 kms northwest of Milan, at the base of the Prealps, close to the Swiss border. The city, which is about 380 m above sea level and serves as a local hub for services and cross-border commuting, is situated between Lake Varese and the surrounding hills, including the Campo dei Fiori Regional Park and the UNESCO Ticino Val Grande Verbano biosphere reserve.

Because of this landscape heritage, Varese has been dubbed the "garden city" [21], highlighting the extraordinary proliferation of villas, parks and gardens that have characterised its urban development over the last three centuries rather than the architectural and urban planning movement that originated in Great Britain at the beginning of the twentieth century. The role played by the bourgeoisie and local entrepreneurs is central to understanding the evolution of the landscape and the widespread recognition of the characteristics of a garden city [22]. The central urban core (the Borgo) had become a temporary residence and favourite destination for some aristocratic Milanese families since the early eighteenth century. From the middle of the following century, however, the rapid development of transport infrastructure, combined with industrial growth, particularly in the cotton industry along the Olona River, and tourism, beginning with the first pilgrimages to the UNESCO site Sacro Monte of Varese and continuing with the construction of large new hotels, became decisive factors in the development of the local economy, influencing the structure of the territory and the development of greenery [23]. Six large parks now public spaces (Giardini Estensi, Ville Mirabello, Baragiola, Torelli Mylius, Augusta, Toeplitz, Castello Masnago and Parco Mantegazza) and the Villa and Collezione Panza, a private place open to the public, bear witness to these significant historical periods.

During the nineteenth century, economic growth strengthened the domination of a class of entrepreneurs who built mansions as holiday residences and places to enjoy the natural landscape. This trend grew rapidly and extensively from the beginning of the twentieth century, a period that saw the consolidation of houses in the Art Nouveau style with extensive private gardens.

The framework of green (from the Sacro Monte to the Olona Valley) and blue networks (from the lake to the hydrographic system), which is recognised as part of the cultural heritage of the Varese area, connects different urban, peri-urban, rural, alpine and lakeside milieus. It reveals a landscape interspersed with vegetation, a part of which is of great historical and environmental value.

In the post-war period, development gradually reduced the vegetation cover, though it remained considerable [24]. At present, more than 80% of the territory is covered by natural or semi-natural areas and agricultural regions, while trees and shrubbery cover almost 60% of the municipal land [25]. The Campo dei Fiori Regional Park is close to the city centre and covers over 5000 ha, including nature reserves and a broad network of hiking trails.

An extensive and diverse vegetation cover is largely a result of the physical and morphological conformation of the territory, which is situated on seven hills, and the landscape, as well as the climatic conditions and the factors associated with its position between the mountains (the Alps), Lake Maggiore and the Milanese plain. The variety of plant species in the urban areas is also linked to romantic gardens, which characterised the development of the garden city: in addition to the impressive conifers (including cedars, araucarias and thuja trees), a huge collection of deciduous trees provides colourful variety and rich foliage.

Despite this strong ecological and landscape legacy, Varese is increasingly affected by environmental vulnerabilities typical of urban systems in transition and by climate change dynamics. Recent increases in soil sealing, especially in central districts with high population density and newly developed areas, have made the city more vulnerable to pluvial flooding, which is brought on by high annual precipitation levels exceeding 1500 mm and inadequate stormwater infiltration. Meanwhile, the Urban Heat Island (UHI) effect has worsened, leading to higher temperatures and thermal discomfort due to a lack of canopy cover and surface sealing, especially in the highly urbanised areas.

These dynamics, combined with habitat fragmentation and the progressive simplification of urban green structures, have caused a significant decline in biodiversity, undermining the city's ecological resilience (Fig. 1).

Fig. 1 Area of study: from the top left, Italy and Lombardy region; on the right, Varese municipality

3 Downzoning and Renaturing Strategy for Varese (10,000)

3.1 Mainstreaming Greening Strategies into the Ordinary Planning Process Through Green and Blue Infrastructure

Renaturing strategies in Varese are guided by the local GBI, which defines the areas whose environmental quality must be preserved. The methodology for developing the GBI is firmly rooted in the ES framework, which serves as a conceptual model for assessing the benefits provided by natural and semi-natural areas with other environmental features. This framework guides the design phase of the GBI, shaping an innovative, climate-adaptive approach to the city's new urban planning agenda.

The GBI was designed considering four different ES through modelling. These are selected according to context-specific conditions of Varese described in the previous chapter, i.e., (i) habitat quality, (ii) stormwater retention capacity, (iii) urban cooling capacity and (iv) sediment retention capacity. The ES assessment identifies the performance of ecosystems in delivering services, as benefits, across the municipality, thereby establishing a solid quantitative basis for understanding the spatial distribution of ecological vulnerabilities and strengths.

The synthesis of the four ES models provides a multisystemic valuation, as detailed in Ghirardelli et al. [26]. This value is enhanced by a qualitative analysis of the interplay between anthropogenic and natural elements shaping the territory and guides the GBI design. This methodological approach introduces a locally adapted strategy that leverages the city's ecological assets while addressing critical environmental challenges, particularly hydrogeological risk, UHI and biodiversity loss. Specifically, the interplay between the natural capital provision and the vulnerability to climate change events informs the conceptualisation of the two fundamental components of the GBI system: structural and adaptive (Fig. 2).

The structural component is the GBI backbone and consists of areas (both public and private) with the highest multisystemic value, encompassing zones identified as biodiversity hotspots. The areas included in this component are categorised into three main classes: (i) urban, comprising urban green spaces (parks, gardens, etc.); (ii) agricultural, consisting of extensively managed farmland with high ecological value; and (iii) natural, represented by forested areas, predominantly located in the northern part of the municipality within Campo dei Fiori Regional Park. Given these areas' high ecological and natural value, the GBI proposes tailored conservation and valorisation strategies, adapted according to targeted areas' land use and spatial characteristics.

The adaptive component consists of public spaces, often intercluded within the dense urban fabric, exhibiting lower multisystemic value. These areas typically display limited biophysical performance due to a combination of factors, including a high percentage of soil sealing, severe morphological conditions, or altered hydrological dynamics. Consequently, these areas exhibit reduced functionality across

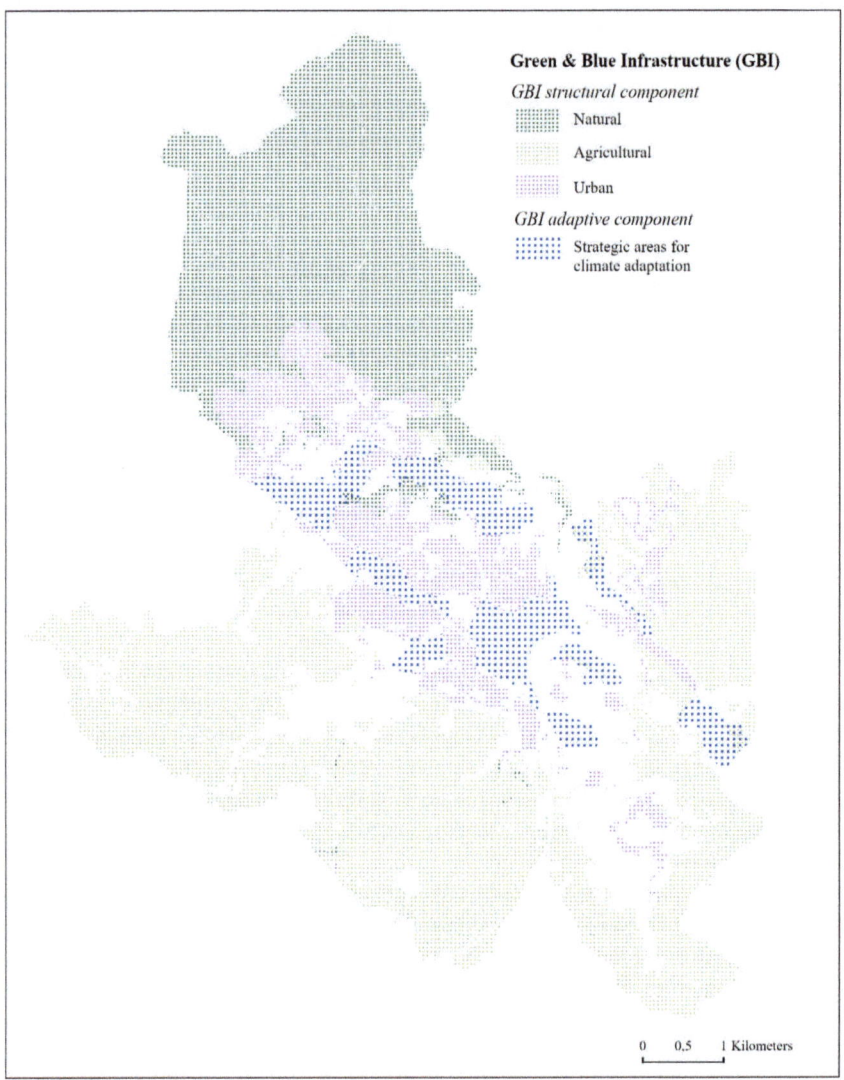

Fig. 2 Varese green and blue infrastructure

three critical ES (stormwater retention, urban cooling and habitat quality), which frequently occur simultaneously. These strategic areas play a key role in climate adaptation (see Fig. 2), therefore, within them, some strategic interventions are required to improve the ES capacity of these areas and reduce the urban vulnerabilities to climate stress. Initially these interventions will be focused on crucial public spaces (streets parking lots and squares) in reducing certain risk factors identifying specific NbS as rain gardens, water square, permeable paving, green roofs, cool materials.

Addressing these areas requires a coordinated, cross-sectoral effort within an integrated, ecologically-informed planning framework. In this context, GBI translates ES assessment into a planning strategy offering a regulatory tool to address structural and adaptive components by defining strategic indications and NbS implementation. The article aims to highlight the operational role of the GBI in supporting policy actions of binding land-take reduction measures with ecological-landscape enhancement and reforestation interventions. This objective defines an experimental framework for testing the implementation tool in achieving NNLT and renaturation targets.

3.2 Downzoning Strategy to Support the GBI Implementation

The regulatory framework within which the GBI action takes place is established by Lombardy Regional Law (RL) no. 31/2014 in fulfilling the EU NNLT target. The legislation establishes objectives, procedures and regulatory mechanisms for governing land take and soil sealing while promoting an urban development that preserves soil functions and incentivises urban regeneration (as explicitly regulated with the RL 18/2019). To reduce land-take, the law sets quantitative limits for local administrations requiring the reduction of area designed for urban transformations in the local plans, as areas, generally natural or agricultural, converted to host human activities (i.e., residential, commercial and tertiary). Specifically, it is requested that urban transformation forecasts not yet implemented be eliminated by reducing or cancelling the building capacity to prevent future land take and preserve the ecological functions of soils. In Varese, the ES-based GBI has allowed defining which areas to include in this downzoning strategy, considering the ecosystem values. So, the downzoning strategy was based on a qualitative assessment of the multisystemic characteristics of each area, allowing for the implementation of tailored-made measures in consideration of ecosystem performance. At the same time, the downzoning strategy supports the local implementation of the GBI by reintegrating downzoned areas into the natural or agricultural system.

The downzoned areas were classified into two main categories accordingly to the multisystemic value and the related strategies proposed:

(i) Areas of Landscape Relevance
(ii) Local Environmental Value Areas

Based on Eggermont et al. [27], for the Areas of Landscape Relevance, measures for protection and restoration are promoted for the preservation and enhancement of their qualities through NBS, such as the maintenance of ecological corridors. Given their large size and strong natural characteristics, they often act as a buffer zone between densely urbanised and more natural or less urbanised areas and are therefore subject to urbanisation pressure. The Areas of Landscape Relevance resulting from downzoning actions cover an area of approximately 20 ha.

Differently, the Local Environmental Value Areas (LEVAs) are smaller in size and exhibit lower, but still significant, multisystemic values. These areas are fundamental for GBI design to enhance their natural value through afforestation and reforestation interventions, which directly benefit the urban areas in which they are located. The downzoning strategy identified 40 LEVAs, covering 6.3 ha of municipal territory ranging from 1500 to 3500 m^2. Their different urban context is also demonstrated by the population density in a 400 m radius, ranging from 190 to 3322 residents. These characteristics highlight the heterogeneity among LEVAs that require GBI strategies tailored to their specific features and local context. Specifically, three different LEVAs conditions are recognised: (i) LEVAs within the GBI, (ii) LEVAs adjacent to GBI and (iii) LEVAs isolated from GBI.

LEVAs within the GBI represent 50% of areas (n = 20). These areas contribute to the GBI structural component, featuring the highest multisystemic values and largest average size (1746 m^2). They are located within natural or rural areas with moderate population density. In this case, LEVAs are subject to interventions for preserving existing ecological functions while strengthening biodiversity provision. Examples of such interventions include: (i) planting native tree clusters to recreate forest structure and canopy diversity; (ii) establishing pollinator meadows using native wildflower seed mixes in open areas; (iii) creating shallow depressions to collect rainwater and support amphibian breeding; and (iv) building brush bundles from pruned native materials along tree lines to provide nesting sites for birds.

The second condition regards LEVAs adjacent to GBI represent 32.5% of areas (n = 13). These areas exhibit intermediate multisystemic values and average size (1229 m^2). They are in transitional zones between urban and natural systems with moderate population density. The areas with medium-higher values are considered part of the GBI's structural component; therefore, the strategy is the conservation of the actual value. For areas with medium-lower values, interventions to enhance the integration of ecological restoration and urban connectivity are proposed. Proposed interventions include: (i) planting multi-row native shrub buffers around perimeters to protect core habitat values; (ii) creating tree corridors using fast-growing native species alongside existing pathways; and (iii) establishing native hedgerows to provide connectivity and wind protection.

The last condition concerns LEVAs isolated from the GBI that represent 17.5% of areas (n = 7). These areas show the lowest current multisystemic values and average size (1670 m^2). They are located within the built-up areas and are subject to significant development pressure. LEVAs are strategic for climate adaptation through intensive NBS and are suitable for increasing urban resilience. Suggested measures include: (i) planting urban forests with climate-resilient native species to maximise the cooling effect; (ii) creating depression areas for rainwater collection, which are lined with native plants that are adapted to periodic flooding and drought; (iii) installing permeable paving with integrated planting strips containing native perennials; and (iv) creating multi-level plant canopies using native trees, understorey shrubs, and ground plants (Fig. 3).

Planning for Biodiversity: Strategies and Actions for Enhancing Nature … 135

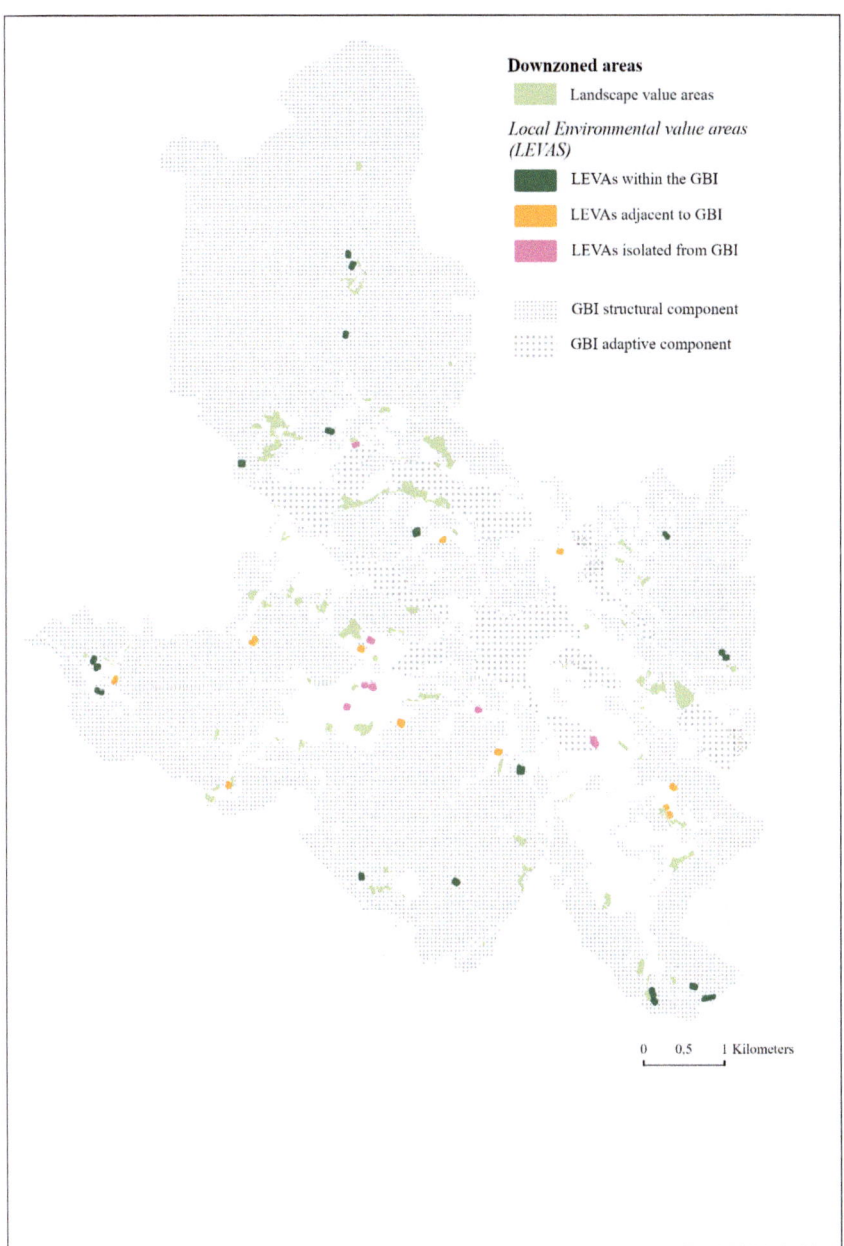

Fig. 3 Downzoned areas categorisation

4 Conclusion

The Varese case study shows how an integrated approach based on ES can be a useful operational tool for coordinating urban planning strategies with the EU's ambitious climate and biodiversity goals. The deployment of a GBI ES-based model provides a replicable model for cities dealing with planning and ecological issues comparable to Varese. Additionally, this approach promotes a more open and evidence-based planning process, which aids local governments in setting priorities for interventions and using public funds best.

The twofold dimension (structural and adaptive) proposed in the GBI enables customised interventions based on vulnerability profiles, urban morphology, and multisystemic value of ES, making it possible to prioritise biodiversity enhancement initiatives where they are most needed while maintaining functional connectivity and spatial coherence throughout the urban landscape. Additionally, this dual structure guarantees planning flexibility across time scales by supporting short-term adaptive responses to climate pressures and long-term conservation goals.

Crucially, this strategic framework incorporates green spaces into a multiscale, multifunctional system that can benefit human health and well-being. The achievement of the NNLT objective depends mainly on the planning regulatory component focused on reducing areas that could determine land-take derived from urban transformation. By reclassifying buildable areas based on ES value, downzoning limits the future land take process and actively reintegrates these areas into the city's GBI, assigning different qualitative targets to each area.

Identifying and classifying LEVAs based on multisystemic indicators and spatial analysis demonstrates how tailored planning actions, ranging from habitat restoration to intensive NBS in dense areas, can strategically contribute to urban regeneration, contrasting future land take processes and increasing urban resilience. Furthermore, the integration between regulatory tools and environmental services capacity represents a fundamental step towards a performance-based and proactive planning in which the measures for protecting and valorising natural capital constitute a fundamental design principle and promote the transition towards more sustainable and biodiversity-rich urban futures.

The case study operationalises ambitious EU policies at the local level, such as the Biodiversity Strategy 2030, the Soil Strategy, and the Nature Restoration Act, providing concrete pathways for cities to become key actors in the ecological transition. As a result, it is necessary and fundamental to overcome the sectoral fragmentation that still characterises many planning systems, integrating the renaturalisation and biodiversity objectives in the traditional planning process and not in sectoral tools or plans. This makes natural strategies more effective as they are embedded in an overall territorial development strategy that considers multiple ecosystem components to which strategies are defined and calibrated in consideration of their respective qualities and performances. In short, the Varese experience serves as a test bed for innovation, offering valuable insights on how ecological performance, regulatory tools and planning practices can work together for a coherent and effective urban strategy.

References

1. IPCC (2021) Climate change 2021: the physical science basis. Contribution of working group I to the sixth assessment report of the intergovernmental panel on climate change. Cambridge University Press, Cambridge, UK; New York, USA
2. Batty M (2008) The size, scale, and shape of cities. Science 319:769–771. https://doi.org/10.1126/SCIENCE.1151419
3. Seto KC, Güneralp B, Hutyra LR (2012) Global forecasts of urban expansion to 2030 and direct impacts on biodiversity and carbon pools. Proc Natl Acad Sci USA 109:16083–16088. https://doi.org/10.1073/PNAS.1211658109/SUPPL_FILE/PNAS.201211658SI.PDF
4. Nilon CH, Aronson MFJ, Cilliers SS et al (2017) Planning for the future of urban biodiversity: a global review of city-scale initiatives. Bioscience 67:332–342. https://doi.org/10.1093/biosci/bix012
5. Fuller RA, Irvine KN (2010) Interactions between people and nature in urban environments. Urban Ecol 134–171. https://doi.org/10.1017/CBO9780511778483.008
6. Brown C, Grant M (1978) Biodiversity and human health: what role for nature in healthy urban planning? Environment 31:326–338
7. European Commission (2019) The European green deal. COM(2019) 640 final. Bruxelles
8. Hermoso V, Carvalho SB, Giakoumi S et al (2022) The EU biodiversity strategy for 2030: opportunities and challenges on the path towards biodiversity recovery. Environ Sci Policy 127:263–271. https://doi.org/10.1016/j.envsci.2021.10.028
9. European Commission (2013) Communication from the commission to the European Parliament, the Council, the European economic and social Committee and the Committee of the regions. Green Infrastructure (GI)—Enhancing Europe's Natural Capital {SWD(2013) 155 final}. Bruxelles
10. European Commission. Directorate-General for Research and Innovation (2015) Towards an EU research and innovation policy agenda for nature-based solutions & re-naturing cities: final report of the Horizon 2020 expert group on "Nature-based solutions and re-naturing cities". Publications Office
11. Dunlop T, Khojasteh D, Cohen-Shacham E et al (2024) The evolution and future of research on Nature-based Solutions to address societal challenges. Commun Earth Environ 5:1–15. https://doi.org/10.1038/s43247-024-01308-8
12. European Commission (2020) EU biodiversity strategy for 2030: bringing nature back into our lives. COM(2020) 380 Final. Brussels
13. European Parliament & European Council (2024) Regulation (EU) 2024/1991 of the European Parliament and of the Council of 24 June 2024 on nature restoration and amending Regulation (EU) 2022/869
14. Marei Viti M, Gkimtsas G, Liquete C et al (2024) Introducing the progress monitoring tools of the EU biodiversity strategy for 2030. Ecol Indic 164. https://doi.org/10.1016/j.ecolind.2024.112147
15. European Commission (2021) EU soil strategy for 2030 reaping the benefits of healthy soils for people, food, nature and climate {SWD(2021) 323 final}. Brussels
16. Kowarik I, Fischer LK, Haase D et al (2025) Promoting urban biodiversity for the benefit of people and nature. Nat Rev Biodivers 214–232. https://doi.org/10.1038/s44358-025-00035-y
17. Runhaar H, Pröbstl F, Heim F et al (2024) Mainstreaming biodiversity targets into sectoral policies and plans: a review from a biodiversity policy integration perspective. Earth Syst Gov 20. https://doi.org/10.1016/j.esg.2024.100209

18. Tardieu L, Hamel P, Viguié V, et al (2021) Are soil sealing indicators sufficient to guide urban planning? Insights from an ecosystem services assessment in the Paris metropolitan area. Environ Res Lett 16. https://doi.org/10.1088/1748-9326/ac24d0
19. Lacoere P, Stefano S, Arcidiacono A (2025) Bridging soil and land targets to reality on the ground. Urban Inf 320:89–90
20. Lacoere P, Leinfelder H (2023) No net land take for Flanders. Towards a roadmap for the implementation of Europe's land target. Raumforsch und Raumordnung/Spat Res Plan 81:620–635. https://doi.org/10.14512/rur.1696
21. Howard E (1902) Garden cities of to-morrow. London
22. Mentasti V (2008) Varese Città giardino. Politecnico di Milano
23. Langé S, Vitali F (1984) Ville della provincia di Varese: Lombardia 2. Rusconi, Milano
24. Cazzola O (1995) Il caso Varese, ascesa e caduta di una città-giardino. Essezeta, Varese
25. Copernicus Land Monitoring Services (2024) Tree cover density 2018—present (raster 10 m), Europe, yearly. https://doi.org/10.2909/e677441e-fb94-431c-b4f9-304f10e4dfd8
26. Ghirardelli F, Mosso B, Ronchi S et al (2024) Integrating ecosystem services performance into urban planning tools: the case of Varese city (Italy). BDC Boll Del Cent Calza Bini 24:63–80
27. Eggermont H, Balian E, Azevedo JMN et al (2015) Nature-based solutions: new influence for environmental management and research in Europe. GAIA Ecol Perspect Sci Soc 24:243–248. https://doi.org/10.14512/GAIA.24.4.9

Open Access This chapter is licensed under the terms of the Creative Commons Attribution-NonCommercial-NoDerivatives 4.0 International License (http://creativecommons.org/licenses/by-nc-nd/4.0/), which permits any noncommercial use, sharing, distribution and reproduction in any medium or format, as long as you give appropriate credit to the original author(s) and the source, provide a link to the Creative Commons license and indicate if you modified the licensed material. You do not have permission under this license to share adapted material derived from this chapter or parts of it.

The images or other third party material in this chapter are included in the chapter's Creative Commons license, unless indicated otherwise in a credit line to the material. If material is not included in the chapter's Creative Commons license and your intended use is not permitted by statutory regulation or exceeds the permitted use, you will need to obtain permission directly from the copyright holder.

Ecological Planning Strategies and Nature-based Solutions in the Context of Climate Change Resilience

Davide Geneletti, Chiara Cortinovis, Chiara Parretta, Simone Caridi, Giuseppe Formetta, Lorenzo Giovannini, Lia Laporta, Alfonso Vitti, and Jarumi Kato-Huerta

Keywords Urban planning · Nature-based solutions (NBS) · Thermal comfort · Stormwater management · Ecosystem services · Spatial modelling

1 Introduction

Climate change is a key challenge for urban areas, where its impacts affect a large number of people and assets. Increasingly frequent and extreme climate events—such as intense rainstorms, heatwaves, and droughts—pose risks to people's health and wellbeing, as well as to both green and grey infrastructure. Cities are thus priority areas for implementing strategies and actions aimed at strengthening climate resilience [1].

Among the available options for strengthening urban climate resilience, nature-based solutions (NBS) have gained traction in recent years, also due to strong endorsement from international institutions. By leveraging ecosystems' processes and functions, NBS can provide multiple co-benefits beyond climate adaptation. For instance, sustainable urban drainage systems and low-impact development techniques can manage stormwater runoff in near-natural ways [2], while street trees and

---

D. Geneletti (✉) · C. Cortinovis · C. Parretta · S. Caridi · L. Laporta · J. Kato-Huerta
PLANES—Planning for Ecosystem Services and Urban Sustainability Lab, Department of Civil, Environmental and Mechanical Engineering, University of Trento, Trento, IT, Italy
e-mail: davide.geneletti@unitn.it

G. Formetta · L. Giovannini · A. Vitti
Department of Civil, Environmental and Mechanical Engineering, University of Trento, Trento, IT, Italy

© The Author(s) 2026
A. De Toni et al. (eds.), *Nature-Positive Cities: Adaptive Spatial Planning in Italy for an Ecological Urban Transition*,
PoliMI SpringerBriefs, https://doi.org/10.1007/978-3-032-06617-6_12

green spaces can cool urban microclimates and reduce the Urban Heat Island (UHI) effect [3]. In addition to mitigating the impacts of a changing climate, these interventions support urban biodiversity and generate positive socio-economic benefits that contribute to the overall resilience of urban areas [4].

Scaling up the use of NBS in urban areas has strong potential for enhancing climate resilience [5, 6]. However, integrating NBS meaningfully into planning processes remains a challenge, largely due to the fragmented nature of urban governance, where multiple sectoral policies and plans operate at different spatial and temporal scales [7]. Spatial planning can play a key coordinating role by ensuring coherent interventions, optimal resource allocation, and equitable distribution of benefits and costs [8]. To be effective, individual NBS must be embedded within broader ecological planning strategies that consider cumulative impacts, synergies, and trade-offs across scales and sectors [9]. This integrated approach also helps bridge the disconnect between short-term interventions and long-term climate resilience objectives [10].

Modelling can be a valid tool to support integrated planning approaches, provided its assumptions and limitations are clearly acknowledged [11]. Modelling enables a deeper understanding of present and future climate-related challenges, allowing for the comparison of alternative scenarios at multiple scales. This is essential for both defining city-wide ecological planning strategies (i.e., at the urban scale) and designing context-specific, locally attuned NBS (i.e., at the local scale). Therefore, models working at different scales and producing useful inputs to inform different decisions must be integrated. Despite the growing recognition of this need in academic research, the practical, operational use of integrated modelling in planning remains limited [12, 13].

This chapter presents recent experience with multi-scale integrated modelling approaches applied in the city of Trento (Italy) to support different planning instruments linked to climate change resilience. Specifically, these include the drafting of Trento's Urban Greening Plan and the preparation of a Masterplan for river restoration. These initiatives were supported by two Horizon-funded projects, SELINA and BioValue, which aimed to integrate ecosystem service knowledge and biodiversity values into spatial planning. At the urban scale, modelling focused on identifying key climate-related challenges, particularly the UHI effect and the risk of urban flooding, and on exploring how they can be addressed through ecological planning strategies. At the local scale, a discussion on the potential use of NBS to enhance urban biodiversity and ecosystem health has sparked interest in exploring its climate-related benefits as well.

The following sections describe the modelling approaches adopted to assess climate-related priorities and inform ecological planning at both urban (Sect. 2) and local scales (Sect. 3). The discussion (Sect. 4) reflects on how the modelling outputs can be used in real-life planning processes, their added value for different policy instruments, and the broader implications for decision-making. We conclude by highlighting limitations and remaining challenges for both research and practice.

2 Defining Ecological Planning Strategies at the Urban Scale

Trento is a medium-sized Alpine city with a peculiar morphology. The most urbanized part of the city lies in the valley floor of the river Adige, where the majority of its 120,000 inhabitants live. Due to its location and the high share of soil sealing, this area is the most prone to urban flooding events. Furthermore, reduced air circulation and a high density of human activities and urban infrastructure make it the most vulnerable to the effects of summer heatwaves, with temperatures often comparable to those of other cities at much lower latitudes. To support the definition of ecological planning strategies at the urban scale, two spatially explicit models were applied to this part of the city (Fig. 1). The first focuses on the risk of flooding and measures runoff generation during an extreme rainfall event, identifying critical areas for stormwater management. The second examines spatial variations in exposure to extreme heat, thus capturing the intensity of the UHI effect.

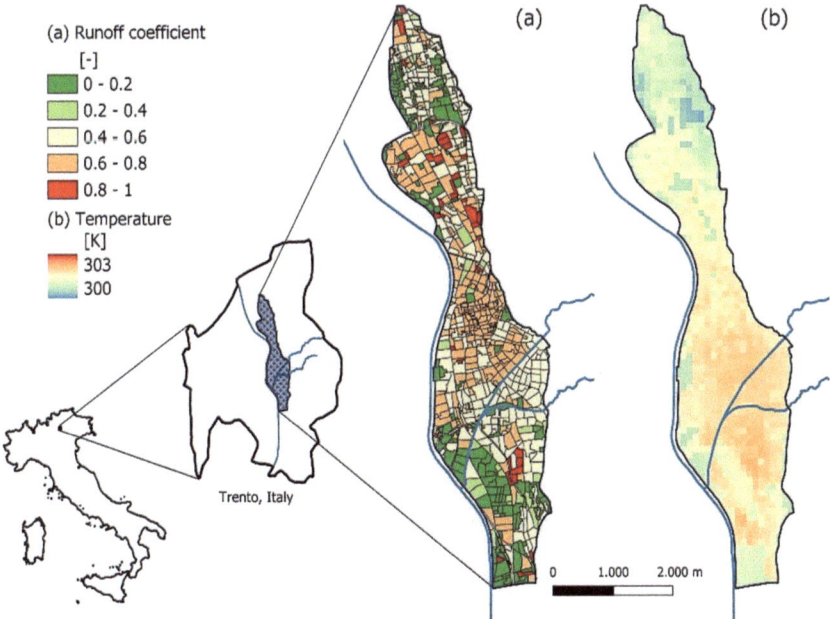

Fig. 1 Location of Trento in Italy and spatial distribution of the Runoff Coefficient (**a**) and average simulated temperature in the period 20–25 August 2023 (**b**) across the valley floor

2.1 Identifying Critical Areas for Stormwater Management

Areas most prone to generate surface runoff were identified for an extreme rainfall event of 15 min with a 100-year return period. The valley floor was divided into sub-catchments, the hydrologic units of analysis, based on the municipal land use map. For each sub-catchment, two complementary metrics were calculated to characterize how rainfall is transformed into runoff. The runoff volume quantifies the absolute amount of water flow generated by each sub-catchment during the selected rainfall event, because of its imperviousness and soil characteristics. This indicator is particularly useful for evaluating the cumulative load on the stormwater system and estimating the potential benefits of volume-reducing measures, such as NBS [14–16]. The Runoff Coefficient (RC) is a dimensionless ratio expressing the fraction of total precipitation that becomes runoff. It is particularly useful for comparing areas with different degrees of imperviousness and identifying zones where interventions to reduce runoff generation, such as green infrastructure or de-sealing, can be prioritized. As such, RC can be applied to identify priority areas for restoration [17] or to estimate the runoff reduction capacity of green infrastructure [18]. At the urban scale, both RC and runoff volume contribute to identifying critical areas for stormwater management and can jointly support the definition of integrated planning strategies [19]. These indicators can be derived using different modelling approaches, from simplified empirical methods (e.g., curve number) to more complex hydrologic models [20].

Figure 1a shows the spatial distribution of RC across the valley floor of Trento. High values (>0.6) are concentrated in the northern industrial-commercial area and in the central urban core, where impervious surfaces dominate. Intermediate values (0.4–0.6) are found in mixed-use zones surrounding the city center. Low values (<0.4) are located in the southern and peripheral northern areas, characterized by agricultural land and green spaces. The observed spatial patterns reflect the degree of imperviousness and provide an evidence-based foundation to support the definition of ecological planning strategies to address the risk of flooding.

2.2 Downscaling Meteorological Models to Assess Exposure to Extreme Heat

A modelling system composed of the Weather Research and Forecasting (WRF) [21] model coupled with a single-layer urban canopy parameterization scheme was used to perform high-resolution forecasts of the thermal field in the urban area of Trento. The WRF forecasts adopt initial and boundary conditions from the Global Forecast System (GFS) model, using three nested domains with a resolution of 9, 3, and 1 km, respectively. WRF forecasts in the inner domain were downscaled to a resolution of 100 m in the urban area of Trento using a single-layer urban canopy model [22], which calculates the energy exchange between the urban environment and the atmosphere, taking into account the physical properties of urban materials and the geometrical characteristics of the city, influencing the surface energy balance and heat trapping inside urban canyons with multiple reflections of longwave radiation. Local characteristics of the urban morphology are accurately taken into account, starting from a detailed land cover map and 1-m resolution Lidar data of building height. The single-layer urban canopy model is offline coupled with WRF, where its meteorological fields at the lowest model level serve as upper boundary conditions.

The modelling chain was applied to simulate meteorological conditions in the period 20–25 August 2023, characterized by a summer heat wave, with minimum temperatures above 20 °C and maximum temperatures above 35 °C in the urban area, reaching almost 40 °C on 23rd August. Figure 1b shows the average simulated temperature field, clearly highlighting the presence of the UHI effect. The temperature field on the valley floor displays a sharp increase at the edge of the city, whereas inside the urban area temperature is closely dependent on the local urban morphology and vegetation fraction. The UHI is stronger during nighttime, with intensities exceeding 2 °C in the core urban area, where the urban morphology is more compact and the vegetation fraction is lower. On the other hand, temperature tends to remain slightly lower in the compact city center during daytime, due to the reduced penetration of solar radiation inside urban canyons and the thermal properties of urban materials, which store heat during the day and release it during nighttime.

3 Designing Nature-based Solutions at the Local Scale

Two analyses were performed at the local scale to support the design of context-specific NBS. The area selected for the local-scale analysis is located on the valley floor, where the Fersina river–subject of the Masterplan for river restoration—flows into the Adige. It includes part of the urban core, as well as vacant lots and public services. The first analysis evaluates the hydraulic performance of the drainage network during an extreme rainfall event; the second analysis simulates the spatial distribution of mean radiant temperature over a hot summer period. Both analyses rely on a detailed map of land cover and vegetation structure characterized using a 30 cm-resolution Pléiades-Neo satellite image from June 2022. In QGIS (v. 3.34.2), non-vegetated and vegetated parcels were identified based on the Brightness Index. A sensitivity analysis was then performed on sample points to determine threshold values for trees and low vegetation [23]. Finally, the results were integrated with data layers on impervious areas, agriculture, and water areas from the municipal land use/land cover map.

3.1 Coupling Hydrologic and Hydraulic Modelling to Select Specific Areas of Intervention

The U.S. Environmental Protection Agency's Stormwater Management Model—SWMM [24] is a widely recognized tool for simulating stormwater runoff in urban areas and is commonly used in stormwater management design and planning (e.g., [25, 26]). It is a dynamic rainfall-runoff model that simulates both the quantity and quality of runoff by representing runoff generation from subcatchments and flow routing through underground drainage systems, which are composed of conduits, nodes, and outfalls [27]. By integrating hydrologic and hydraulic components, the model evaluates surface water flooding by simultaneously assessing runoff production and its downstream impacts on the drainage network.

Here we employed two modeling approaches provided in SWMM: (i) the SCS-CN method [28] for simulating spatially variable infiltration and runoff generation, and (ii) dynamic wave routing for modeling flow movement through the drainage system. The SCS-CN method accounts for variations in soil type and land use/land cover, which influence runoff production. We used high-resolution land cover data to define imperviousness and infiltration parameters for each sub-catchment. The dynamic wave routing method (e.g., [24]) captures complex hydraulic phenomena such as pressurized flow, channel storage, flow reversal, backwater effects, and entrance/exit losses.

The study area was divided into 80 sub-catchments with 76 junctions and 73 pipes. Each sub-catchment drains stormwater to its nearest junction. The drainage network of the study area was initially sized using a design rainfall event with a 15-min duration and a 10-year return period, ensuring that conduits operate at no more than 80% of their full capacity. Subsequently, the system was evaluated under a more extreme pluvial event of the same duration (15 min) but with a 100-year return period. Following [29], the hydraulic impact of this event was assessed in terms of node flooding and conduits' degree of fullness. Two synthetic indicators were derived from SWMM model outputs: (i) number of flooded nodes (defined as nodes with a flood volume greater than zero), representing the system's ability to manage inflows; (ii) conduit degree of filling, calculated as the ratio of the maximum water depth to the full depth of each conduit, indicating the extent of hydraulic loading.

The results reveal critical areas of the network experiencing significant hydraulic stress under extreme conditions. As shown in Fig. 2a, 60% of conduits exceed a filling degree of 0.8, with the most affected segments located in the downstream portion of the system. These conduits are particularly vulnerable to surcharging, primarily due to the accumulation of upstream flows. Additionally, 14% of nodes are identified as flooded, corresponding to junctions where critical conduits converge. These locations are prone to localized overflows, driven by limited drainage capacity and high internal pressure in the conduits.

Overall, the findings pinpoint sections of the drainage network where the system is likely unable to efficiently manage stormwater during extreme rainfall scenarios (conduits and nodes marked in red in Fig. 2a), underscoring the need for targeted

Fig. 2 Spatial distribution of flooded nodes and conduit degree of filling of the drainage network (**a**), and average simulated mean radiant temperature within the study area in summer 2023 (**b**)

mitigation measures to reduce runoff volumes and providing preliminary insights for NBS prioritization.

3.2 Modelling Thermal Comfort to Inform the Strategic Siting of NBS

Urban Multi-scale Environmental Predictor (UMEP) is a modular open-source QGIS extension developed to support urban microclimate analysis [30]. Among its modules, SOLWEIG (SOlar and LongWave Environmental Irradiance Geometry) is used to estimate mean radiant temperature (Tmrt) [31, 32], a comprehensive indicator of the radiative environment perceived by the human body, in contrast to air temperature, which is independent of radiative influence [33].

The model simulates a 2.5D urban environment through 1-m-resolution DEM, DSM, and land cover layers. It accounts for all visible surfaces emitting thermal radiation, including both direct and diffuse solar radiation as well as long-wave radiation emitted from buildings, ground surfaces, and vegetation [34], while also incorporating the effects of shading and multiple reflections. SOLWEIG integrates hourly meteorological data (ERA5, Copernicus Climate Change Service). These inputs enable fine-scale microclimate simulations and the production of high-resolution hourly maps of Tmrt (Fig. 2).

During the summer of 2023, average Tmrt values in the study area ranged from 18.6 to 27.8 °C. When considering only daytime hours, the average Tmrt reached up to 38.9 °C, with maximum hourly peaks exceeding 68 °C in the most exposed locations. The critical threshold of 55.5 °C, associated with a 5% increase in heat-related mortality risk for individuals over the age of 80 [33], was exceeded for almost 0.2% of the total daytime in several parts of the study area. These exceedances were primarily associated with high levels of solar exposure, the prevalence of impervious surfaces, and the limited presence of vegetation and shaded areas, all of which reduce the potential for radiative and evaporative cooling. Additionally, urban morphological features such as low sky view factor, the prevalence of heat-retaining materials, and the lack of tree canopy contribute to further exacerbating the radiative heat load on individuals.

In addition to its standalone relevance for assessing heat-related health risks, Tmrt plays a critical role in estimating the Universal Thermal Climate Index (UTCI), as it reflects the actual radiative thermal load on the human body. UTCI is an advanced biometeorological index that estimates the apparent temperature perceived by the human body based on air temperature, relative humidity, wind speed, and Tmrt [35]. Two points of interest (PoIs) were identified within the study area: the first (PoI 1) is located in an exposed urban setting characterized by impervious ground surfaces and surrounded by buildings, while the second (PoI 2) is situated within a residential green space, featuring trees and grass (see Fig. 2b). For both locations, Tmrt and UTCI values were calculated for a standing 80-year-old male subject over the period from 20 to 25 August 2023 (the same time frame used for the temperature modelling in Fig. 1b). The results are presented in Fig. 3. The Tmrt values observed at PoI 1 regularly exceeded the critical threshold of 55.5 °C, with frequent peaks above 60 °C. In contrast, Tmrt at PoI 2 consistently remained below the critical threshold throughout the observation period, confirming the moderating influence of vegetation evapotranspiration and shading on the local radiative environment. Regarding thermal comfort, UTCI values at PoI 1 generally exceeded 26 °C (a level typically associated with moderate heat stress) and reached values above 32 °C (strong heat stress) during the peak hours (12:00 and 14:00). At PoI 2, the overall diurnal pattern was similarly shaped, but the peak values were markedly lower, exceeding the 32 °C threshold only in isolated instances.

4 Discussion and Conclusions

The models applied in Trento, targeting urban flooding and thermal comfort, produced decision-relevant outputs at both urban and local scales.

At the urban scale, the two models applied to the most climate-vulnerable area of the city, i.e., the valley floor, provide valuable insights into the identification of the priority areas that ecological planning strategies should target. The runoff

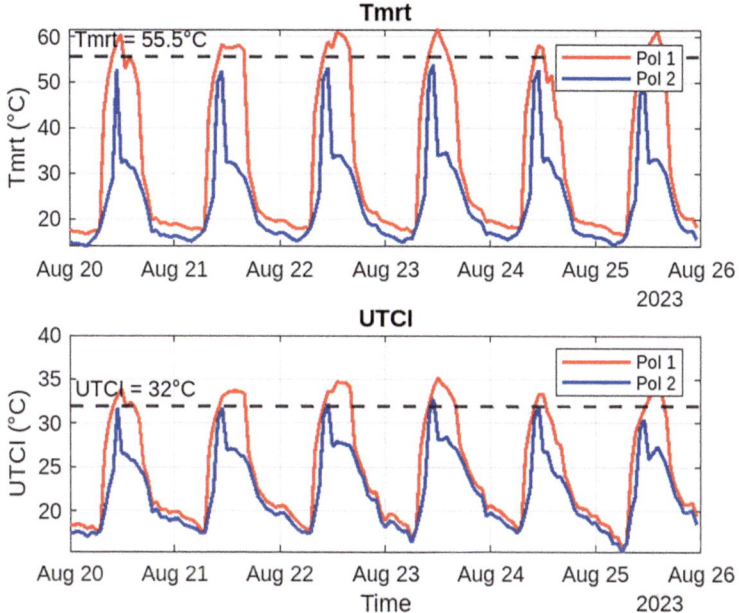

Fig. 3 Temporal evolution of Tmrt and UTCI at two selected points of interest: PoI 1 = impervious surfaces surrounded by buildings, PoI 2 = residential green space with tall trees and grassy areas

coefficient highlights sub-catchments that, due to their high degree of imperviousness, have limited capacity to infiltrate water. These areas produce relatively more runoff, which is ultimately responsible for urban flooding [36]. Similarly, the map of air temperature identifies areas of the city where temperatures are higher than those of the surroundings, due to their morphological and vegetational characteristics. Drawing on this knowledge, ecological planning strategies can guide actions to improve stormwater management, such as removing impervious surfaces or installing green roofs, primarily in industrial and commercial zones in the northern part of the city. Simultaneously, actions to enhance the local microclimate, such as tree planting, should preferably be directed towards the city center and the southern neighborhoods.

While models applied at the urban scale can inform strategic priorities, higher resolution models using more detailed inputs can be applied to guide the design of site-specific NBS. For instance, the two models run at the local scale served to identify locations where the implementation of NBS could be more beneficial for climate resilience. Crucially, higher spatial resolution enhances the capacity of models to capture the fine-grained spatial variability that often exists within broader priority areas. Compared to the use of models at the urban scale, these more localized applications not only relied on higher resolution input data to produce more detailed outputs, but also allowed for the simulation of more complex processes. For example, SWMM combines the analysis of the hydrologic processes described by the runoff coefficient with hydraulic modelling of the drainage network's performance.

This allows assessing the impact of proposed NBS not only in terms of reduced runoff, but also through other indicators more directly linked to NBS benefits, such as the reduction of flood volume (when nodes are flooded) or flood risk (associated with the conduits' degree of filling). Similarly, the thermal indices estimated using the UMEP model, such as mean radiant temperature and the Universal Thermal Climate Index, capture the level of thermal stress experienced by people in outdoor spaces. Unlike air temperature, these user-centered indicators are well-suited for evaluating the benefits of NBS from a human comfort perspective, thus being more informative for designing and assessing the potential impacts of NBS.

In practice, the models applied at both urban and local scales in Trento offered valuable inputs for different types of planning instruments. At the city scale, model results informed the development of the Urban Greening Plan. The plan responds to the objectives set by the EU Biodiversity Strategy for 2030, which calls on cities of at least 20,000 inhabitants to "develop ambitious Urban Greening Plans" (now "Urban Nature Plans") including "measures to create biodiverse and accessible urban forests, parks and gardens; urban farms; green roofs and walls; treelined streets; urban meadows; and urban hedges" [37]. Trento's Urban Greening Plan positions climate change as a central challenge to address through greening interventions and includes spatially-explicit modelling outputs, particularly on runoff generation and urban microclimate, to inform future greening and ecological planning strategies and interventions. These outputs can be used to prioritize specific types of NBS in different parts of the city, hence supporting a more targeted and potentially effective response. Modelling results further support the argument that the multifunctionality of urban green infrastructures and the synergies among multiple ecosystem services should be acknowledged and strategically leveraged in the design of ecological strategies for climate-resilient cities [10].

To be implemented at the local scale, such ecological planning strategies require the detailed design of context-specific NBS. This implies a preliminary analysis of the potential impacts of different solutions, to support decisions about their location and specific design characteristics (e.g., size, materials, species) [38, 39]. These potential impacts must be evaluated in conjunction with other parameters that define the feasibility of the proposed interventions, including the availability of space and resources [40]. In Trento, the ongoing development of a Masterplan for the restoration of the river Fersina offered the opportunity to reflect not only on the feasibility of specific interventions, but also on their benefits.

The limitations related to the use of the described models should also be acknowledged. Although we do not address the technical limitations of each model here (for which we refer to the relevant literature), we highlight key considerations regarding their use in policy processes. First, there is, in general, a trade-off between simplicity and accuracy, meaning that the more accurate a model is, the higher the risk that it is perceived by the actors involved in the process as a black box or an all-solving "Oracle" [41]. In the described applications, however, we have also observed that the complex models applied at the local scale produce indicators that might be more meaningful to a non-technical audience, as they focus on variables directly linked to people's perceptions. This suggests that when selecting the most appropriate model to use, it is essential to consider not only its assumptions, but also its outputs and how both can be effectively communicated to the actors involved in the process.

Second, the breadth and depth of analysis should be balanced by considering the type of decisions being supported. Ecological planning strategies and NBS are multipurpose and, beyond addressing the main challenge for which they are implemented, they produce several co-benefits. These co-benefits are rarely captured by modelling approaches such as the ones described above. Yet at the strategic level, it is especially important to consider and quantify them not only to justify the adoption of NBS over other (grey) solutions, but also to anticipate additional consequences that will emerge from the implementation of the proposed interventions, especially in terms of spatial equity and distributive justice [42]. For instance, many planning processes fail to assess how the benefits and burdens of NBS are distributed across different social groups in urban areas. This can lead to interventions that, while ecologically effective, reinforce existing spatial injustices [43]. As such, it is important to combine multiple assessments, which often implies a trade-off with the accuracy of analysis.

Finally, a multiscale approach, such as the one presented in this chapter, involves challenges in the application of the results in real-life decision-making processes. While having different types of indicators at different scales makes sense, it also implies that the results will not perfectly overlap across scales. Managing these inconsistencies and communicating the implications to stakeholders requires careful expectation management and a conscious effort to avoid overly technocratic interpretations of model results. As observed in similar planning contexts [44], the role of experts in mediating between modelling outputs and planning processes is crucial to ensure transparency, legitimacy, and shared understanding of results.

Acknowledgements The authors acknowledge support from SELINA and BioValue projects. The projects receive funding from the European Union's Horizon Europe research and innovation programme under grant agreements No 101060415 and No 101060790, respectively. Views and opinions expressed are however those of the authors only and do not necessarily reflect those of the European Union or REA. Neither the European Union nor the granting authority can be held responsible for them.

References

1. Reckien D, Buzasi A, Olazabal M, Spyridaki NA, Eckersley P et al (2023) Quality of urban climate adaptation plans over time. npj Urban Sustain 3:13
2. Zahmatkesh Z, Burian SJ, Karamouz M, Tavakol-Davani H, Goharian E (2015) Low-impact development practices to mitigate climate change effects on urban stormwater runoff: case study of New York City. J Irrig Drain Eng 141(1):04014043
3. Marando F, Heris MP, Zulian G, Udías A, Mentaschi L et al (2022) Urban heat island mitigation by green infrastructure in European Functional Urban Areas. Sustain Cities Soc 77:103564
4. Faivre N, Fritz M, Freitas T, De Boissezon B, Vandewoestijne S (2017) Nature-Based Solutions in the EU: innovating with nature to address social, economic and environmental challenges. Environ Res 159:509–518

5. Cortinovis C, Olsson P, Boke-Olén N, Hedlund K (2022) Scaling up nature-based solutions for climate-change adaptation: potential and benefits in three European cities. Urban For Urban Greening 67:127450
6. Orta-Ortiz MS, Geneletti D (2023) Prioritizing urban nature-based solutions to support scaling-out strategies: a case study in Las Palmas de Gran Canaria. Environ Impact Assess Rev 102:107158
7. Martin J, Scolobig A, Linnerooth-Bayer J, Irshaid J, Aguilera Rodríguez JJ et al (2025) The nature-based solution implementation gap: a review of governance barriers and enablers. J Environ Manage 388:Article 126007
8. Corgo J, Cruz SS, Conceição P (2024) Nature-based solutions in spatial planning and policies for climate change adaptation: a literature review. Ambio 53(11):1599–1617
9. Geneletti D, Cortinovis C, Zardo L, Adem Esmail B (2020) Planning for ecosystem services in cities. Springer International Publishing
10. Bush J, Doyon A (2019) Building urban resilience with nature-based solutions: how can urban planning contribute? Cities 95:102–113
11. Walther F, Barton DN, Schwaab J et al (2025) Uncertainties in ecosystem services assessments and their implications for decision support—a semi-systematic literature review. Ecosyst Serv 73:101714
12. Verburg PH et al (2008) Land system change and food security: towards multi-scale land system solutions. Curr Opin Environ Sustain 1(3):219–225
13. Lim B et al (2017) Integrating climate change risks into urban planning: a review of tools and approaches. Sustainability 9(3):523
14. Selbig WR, Loheide SP, Shuster W, Scharenbroch BC, Coville RC et al (2022) Quantifying the stormwater runoff volume reduction benefits of urban street tree canopy. Sci Total Environ 806(3):151296
15. Gao J, Wang R, Huang J, Liu M (2015) Application of BMP to urban runoff control using SUSTAIN model: case study in an industrial area. Ecol Model
16. Lähde E, Dahlberg N, Piirainen P, Rehunen A (2023) Ensuring ecosystem service provision of urban water nature-based solutions in infill areas: comparing Green Factor for districts and SWMM modeling in scenario assessment. Environ Process
17. Soto-Montes-de-Oca G, Cruz-Bello GM, Bark RH (2023) Enhancing megacities' resilience to flood hazard through peri-urban nature-based solutions: evidence from Mexico City. Cities 143:104571
18. Li L, Van Eetvelde V, Cheng X, Uyttenhove P (2020) Assessing stormwater runoff reduction capacity of existing green infrastructure in the city of Ghent. Int J Sust Dev World 27(8):749–761
19. Zhai J, Ren J, Xi M, Tang X, Zhang Y (2021) Multiscale watershed landscape infrastructure: integrated system design for sponge city development. Urban For Urban Greening 60:127060
20. Kumar S, Vishwakarma RK, Tyagi VK, Kumar V, Kazmi AA, Ghosh NC et al (2024) Stormwater runoff characterization and adaptation of best management practices under urbanization and climate change scenarios. J Hydrol 635:131231
21. Skamarock WC, Klemp J, Dudhia J, Gill D, Liu Z, Berner J et al (2021) A description of the advanced research WRF model version 4.1. NCAR technical notes NCAR/TN-556+STR
22. Giovannini L, Zardi D, de Franceschi M (2013) Characterization of the thermal structure inside an urban canyon: field measurements and validation of a simple model. J Appl Meteorol Climatol 52:64–81
23. Qian Y, Zhou W, Nytch CJ, Han L, Li Z (2020) A new index to differentiate tree and grass based on high resolution image and object-based methods. Urban For Urban Greening 53:126661
24. Rossman LA (2015) Storm water management model user's manual version 5.1. U.S. Environmental Protection Agency
25. Zhou Q (2014) A review of sustainable urban drainage systems considering the climate change and urbanization impacts. Water 6(4):976–992
26. Nanda AR, Nurnawaty Mansida A, Bancong H (2025) A bibliometric analysis of trends in rainfall-runoff modeling techniques for urban flood mitigation (2005–2024). Results Eng 26:104927

27. Rossman LA, Simon MA (2022) Storm water management model user's manual version 5.2. U.S. Environmental Protection Agency
28. USDA (1986) Urban hydrology for small watersheds. United States Department of Agriculture, Washington, DC, Technical Release 55 (TR-55)
29. Zhang Y, Zhao W, Chen X, Jun C, Hao J, Tang X, Zhai J (2021) Assessment on the effectiveness of urban stormwater management. Water 13(1):4
30. Lindberg F, Grimmond CSB, Gabey A, Huang B, Kent CW et al (2018) Urban Multi-scale Environmental Predictor (UMEP): an integrated tool for city-based climate services. Environ Model Softw 99:70–87
31. Lindberg F, Holmer B, Thorsson S (2018) SOLWEIG 2018a—a model for calculating the mean radiant temperature in complex urban settings. Urban Clim 24:1–20
32. Lindberg F, Grimmond CSB (2011) The influence of vegetation and building morphology on shadow patterns and mean radiant temperatures in urban areas: model development and evaluation. Theoret Appl Climatol 105:311–323
33. Thorsson S, Rocklöv J, Konarska J, Lindberg F, Holmer B, Dousset B, Rayner D (2014) Mean radiant temperature–a predictor of heat related mortality. Urban Clim 10:332–345
34. Thorsson S, Lindberg F, Eliasson I, Holmer B (2007) Different methods for estimating the mean radiant temperature in an outdoor urban setting. Int J Climatol 27(14):1983–1994
35. Jendritzky G, De Dear R, Havenith G (2012) UTCI—why another thermal index? Int J Biometeorol 56:421–428
36. Guo K, Guan M, Yu D (2021) Urban surface water flood modelling—a comprehensive review of current models and future challenges. Hydrol Earth Syst Sci 25:2843–2860
37. European Commission (2020) EU biodiversity strategy for 2030. COM(2020) 380
38. Cortinovis C, Geneletti D (2019) A framework to explore the effects of urban planning decisions on regulating ecosystem services in cities. Ecosyst Serv 38:100946
39. Orta-Ortiz MS, Geneletti D (2022) What variables matter when designing nature-based solutions for stormwater management? A review of impacts on ecosystem services. Environ Impact Assess Rev 95:106802
40. Longato D, Cortinovis C, Balzan M, Geneletti D (2023) A method to prioritize and allocate nature-based solutions in urban areas based on ecosystem service demand. Landsc Urban Plan 235:104743
41. Townsend A (2015) Cities of data: examining the new urban science. Public Culture 27(2)
42. Ommer J, Bucchignani E, Leo LS et al (2022) Quantifying co-benefits and disbenefits of nature-based solutions targeting disaster risk reduction. Int J Disaster Risk Reduction 75:102966
43. Kato-Huerta J, Geneletti D (2022) Environmental justice implications of nature-based solutions in urban areas: a systematic review of approaches, indicators, and outcomes. Environ Sci Policy 138(July):122–133
44. Webber MK, Samaras C (2022) A review of decision making under deep uncertainty applications using green infrastructure for flood management. Earth's Future 10(7):e2021EF002322

Open Access This chapter is licensed under the terms of the Creative Commons Attribution-NonCommercial-NoDerivatives 4.0 International License (http://creativecommons.org/licenses/by-nc-nd/4.0/), which permits any noncommercial use, sharing, distribution and reproduction in any medium or format, as long as you give appropriate credit to the original author(s) and the source, provide a link to the Creative Commons license and indicate if you modified the licensed material. You do not have permission under this license to share adapted material derived from this chapter or parts of it.

The images or other third party material in this chapter are included in the chapter's Creative Commons license, unless indicated otherwise in a credit line to the material. If material is not included in the chapter's Creative Commons license and your intended use is not permitted by statutory regulation or exceeds the permitted use, you will need to obtain permission directly from the copyright holder.

MIX
Papier aus verantwortungsvollen Quellen
Paper from responsible sources
FSC® C105338

If you have any concerns about our products,
you can contact us on
ProductSafety@springernature.com

In case Publisher is established outside the EU,
the EU authorized representative is:
**Springer Nature Customer Service Center GmbH
Europaplatz 3, 69115 Heidelberg, Germany**

Printed by Libri Plureos GmbH
in Hamburg, Germany